AF683976

# THE HOUSE THAT JACKIE BUILT

a simple guide to
building in stone
with formwork

by Jackie French

illustrations by Bryan Sullivan

EARTH GARDEN BOOKS
Trentham, Victoria

Published by EARTH GARDEN BOOKS
PO Box 2
Trentham,
Victoria, 3458,
AUSTRALIA.
Phone (03) 5424 1819 Fax (03) 5424 1743.
Email: <info@earthgarden.com.au>.
World Wide Web site: <www.earthgarden.com.au>
A.C.N. 086 043 567
A.B.N. 69 086 043 567

© Copyright Jackie French 2001

ISBN 0 9586397 8 6

All rights reserved. No part of this publication may be reproduced, stored in a retrieval system, or transmitted in any form or by any means, electronic, mechanical, photocopying, recording, or otherwise, without the express permission of the publishers, except in the case of brief quotation embodied in critical articles and review.

This book is printed on plantation-grown paper in the interests of preserving our planet's precious resources.

Edited by Alan T Gray
Editorial Co-ordinator: Linda Wheeler
Graphic design, index, all-round brilliance: Tony Fuery

Printed by McPherson's Printing Group, Victoria.
Distributed by Gordon and Gotch, and Gemcraft Books, Melbourne.

DISCLAIMER
Due to the variability of local conditions, materials, skills, site and other such factors, Earth Garden Publishing Pty Ltd assumes no liability for personal injury, property damage, or loss of any kind whatsoever arising from actions inspired by information in this book.
The publishers and author cannot accept responsibility for any errors or omissions, or for any claim arising from the use of the book or any information contained in it.
The information contained in this book is of a general nature, and is not intended to be a substitute for professional advice.

# CONTENTS

*to Bryan, and all the others who built this house*

# INTRODUCTION

## Two apologies and an explanation

First of all I have to apologise for the name. I DON'T live in 'the house that Jackie built'!

I live in the house that Jackie, Bryan, Ray, Steve, Hugh, David, Val, Libby, Noel, Geoff, Michael, David (another one) Geoff (another one) Richard, Clements, Cindy, Jane and lots of other people built too, including a passing clergyman on a pastoral visit who was nearly knocked unconscious as he helped us raise a roof beam. He didn't call again, and as the house slowly keeps growing more people will undoubtedly be added to that list.

But I have been part of every stage of the building; mixed the concrete, gathered the rocks, built stone walls . . . and it's the bit that I was most part of that I am writing about now: how to build slipform stone walls, a technique that even I can manage.

Which brings me to the second apology.

I'm not a builder, or a carpenter (can't even drive a nail in straight most of the time) or an engineer or an architect, which means any advice you get from me is probably at least partly uninformed and incompetent and I won't explain things like a professional might.

But I did build a large part of my own house — and it's stayed upright — and in this book I'm simply telling you how I (and many others) did it.

Please forgive any lack of professionalism in this book — it's unavoidable. On the other hand, if I can do it, ANYONE can and, as an amateur, perhaps I can explain things for other amateurs in a way a professional mightn't. I hope so anyway.

Building this place was fun — still is fun. It was also surprisingly cheap and above all POSSIBLE. If I can communicate the last three things to you I'll have done what I wanted — and you can always consult professionals for advice (see below).

### Disclaimer

This book should not be used in place of professional engineering, building, legal or architectural advice. ALL stone building is dangerous. Rock is heavy; accidents happen. Build at your own risk!

### Getting Permission

You <u>must</u> get a building permit from your local council before you build a house or various other structures. Most councils will require you to get an engineer's certificate before they'll give you permission to build a stone house. This is in case it falls down. If it does, you might sue the council . . . and if you do, the council might be able to sue the engineer who said it'd work.

Actually, do discuss your plans with an engineer anyway, before you finalise them. The size (thickness and length and height) of your walls, your lintels, your foundations and the amount of reinforcing you need will all vary enormously according to the size and dimensions of your house, your soil type, rock type and house site. If you get these wrong you may end up with a disaster! <u>All</u> advice in this book is a guide only — you must be <u>sure</u> your design is safe before you start!

An engineer will be able to give you answers that I'm simply not competent to give . . . and couldn't give, anyway, without knowing what you plan.

# ONE

# The Story of Our House

THE story of most houses is simple: the land is bought by a developer wanting to make a profit; the house is built as cheaply as possible, with no real thought about who is going to live there, or what sort of house they need.

It's bought with a mortgage and lived in for a few years till the inhabitants save up enough for an upgrade. When I say 'lived in' I mean mostly inhabited at night and weekends, with holidays spent at the Gold Coast, and even the time that is actually spent in the house is mostly occupied by sleeping or TV.

This house is different.

This is a house that is lived in. We work here, laugh and talk here, discuss the world and meet our friends here. It's a house that is used 24 hours of just about every day — and it's designed to accommodate all that living. (We don't go on holiday much . . . well, never actually. What do we need a holiday from? I'd rather explore the cliffs above the house, hunt for last year's eagle's nest, or float in the creek and watch the sky and casuarinas. Besides, I might miss something . . . the first of the roses unfolding, the last navel orange before the bower birds pinch it . . . ).

The story of this house is the story of my life for the past 25 years, with other people's stories too, because I'm only one of many people who have made this house what it is, though I suppose I'm the one who has lived in it longest, and been part of all its phases (beating my son in the longest occupation stakes by eighteen months).

A lot of laughter and love and happiness went into building this house. I don't think a week goes by without my catching sight of a certain stone or lintel and remembering — the reason that wall bulged is because it was so hot on the day we placed it in the wall, we finished it fast and dashed down to the creek for a swim; that's the good flat rock that Noel found with what's his name up the gorge (not her present partner so I'll be discreet).

That's the bit of wall with the gold nugget in it — I meant to put it to one side but it was used by mistake; and that's the wall that Libby and Jane and I built while taking turns at reading aloud from a 1950s *Women's Weekly* book of etiquette that we found at the dump (I learnt you never ever ask the Queen a direct question. If you want to say: "Where's the loo, Liz?" replace it with: "I assume the toilet facilities are to the left," and she'll graciously reply, "No, dear subject, they are to the right." If I hadn't built my own house I never would have known that . . . ).

In between we sang all the hymns we could remember from our childhoods and competed to find the most ludicrous popular song of all. *As Long as He Needs Me* from *Oliver* was a particular favourite, if I remember rightly, followed by *He May Not Always Say What You Would Want Him To* from *The King and I*. Every stone of this house has memories. You are incredibly rich in memories when you live in a house you built yourself.

I don't have the right to tell the stories of the other people who built this house, except where their stories are part of mine, and even then there are things it wouldn't be fair to say. This is just my part of the story. I suppose any owner built house is a complex of many people's stories too.

There are some bad memories that go with this house as well: the break up of my first marriage, the times of loneliness and poverty when Edward was a baby, washing the nappies by hand and reading by gas light — except it really didn't seem so bad at the time, and there was a young joey called Fuchsia for company and while washing the nappies I counted the dropped foodstuffs of the polliste wasps that nested above the sink and so made my first really long term study of who ate what pest here . . . and anyway you don't really feel poor with a creek and a garden and wombats at the back door — at least I didn't.

The only real terror of poverty was that I might not be able to afford to buy out my ex-husband's share of the property to be able to keep living here.

**The house is part of the garden and the garden is part of the house.**

The story of this house contains most of my adult life — but it can't just be a story of a house and its people either, because the house is also part of the garden. Or is the garden part of the house?

And the garden is part of the wider ecology of this end of the valley, and the valley has been the basis of my life's work too, whether it was growing fruit and vegies to sell years ago, or writing about gardening and pest control or writing the fiction that is all really based here, in this community, this valley, this garden, this house . . .

It's fashionable nowadays to accept the Aboriginal bond with the land. I think most people though would think it strange if I said that I am part of this house, this garden, the hills and gorge around it. They have changed me far more profoundly than I have changed them. This is my country. There is no separating us; and when I die I hope I stay part of it too.

## Why build your own?

Well, the first reason, of course, is cost. That's why most of us decide to build our own houses. It's possible to put up the shell of a house for about $5,000.

Don't think you're going to automatically have a cheap house if you build the walls from stone though — or from mudbrick or rammed earth or any other cheap 'alternative'.

Most of the cost of a house isn't the basic structure — it's the plumbing, the kitchen cupboards, floor coverings, stove, electricity connection or alternative power system . . . but a small basic house can still be built for about $10,000 to $12,000. The great advantage of a home-built house — like a home-built power system — is that you can add to it as you have the money.

## Houses that grow

Instead of living under the burden of a $100,000 loan most of your life (and the even greater burden of interest bills — you'll have paid many times that by the time you're finished) you can save up and buy your land outright; save up another $3,000 or so and put in foundations; save up another couple of thousand and put up framework and roof, and while you're spending your weekends and holidays build-

### NOT JUST SAVING MONEY

*I'd keep chooks even if they never gave me any eggs . . . for their manure, but also for their cluck, and their personalities. I've never felt my chooks have to pay for themselves (though I'm sure they do). Same with houses. Okay, it's cheaper to build your own, but that isn't the main reason for doing it.*

*The more you put into a place, the greater your connection with it. I always feel that if you just plonk down money for a house — no matter how lovely — you must always have a sense you're just camping there. Building the house that you will live in: actually designing it, building it stone by stone, planting the garden around it and watching both grow from nothing — is one of the most creative, satisfying things you can do.*

ing you're saving for the next bit.

You move into the first bit of it, and keep adding more bits . . . another bedroom or two or three, another bathroom perhaps, a living room or study . . . as you need them and can afford them, doing at least most of the work yourself.

Our house, like that of many owner builders, has been built in stages — a one bedroom stage, a two bedroom . . . then we added an upstairs, then a study, then enlarged the living room . . . all done when we had the money to do it, so we didn't have to borrow.

This lack of debt has given us a real sense of freedom. Okay, it's impossible to be totally self-sufficient — there are still rates to pay — and anyone with a potential teenager in the household is kidding themselves if they think that their offspring will necessarily be happy to live a totally different life from their friends.

But with no mortgage and a garden of fruit, vegies and useful plants we have had a lot more freedom to do the sort of work we want to, not to mention spending the odd hour walking and swimming if we feel like it.

How do you save the initial money? To be honest, I'm horrified at the amount of money most people in their twenties go through. At the risk of sounding like the cliche: "It was different in my day . . .". Well, it was.

Nowadays most young couples assume they'll rent a flat of their own; most singles share with only one or two others. Rent eats up a good third of their budget.

Forget about a home of your own till it IS your own. Share a large house with lots of people; as soon as possible move into a caravan on your land (you can sell the caravan afterwards.) Or into a shed if you can get council permission.

A friend of ours bought her own house last year — a tiny one that needs much renovating and adding to. She's a single mother with two adolescent kids, living mostly on the supporting parent's benefit with a few odd cleaning jobs to supplement it. (The kids do odd jobs too.)

She saved about $2,500 a year — but she's a whizz at saving money and a whizz at luxuries that don't cost much — a picnic at the river once a week instead of eating out, homemade hamburgers and homemade popcorn and video every Friday night, garage sale bargains. Actually, it's not the amount of money she saves that is so fascinating — it's the amount of fun her family has while she does it.

Any couple who are both working should be able to save enough to buy their land in five years, and with luck— and a cheap block— it should take

## GIVING SOMETHING TO THE FUTURE

*All that we have today our ancestors bequeathed to us: the still unexploited natural resources, our schooling, our literature, the house we grew up in, the public buildings. I sometimes wonder if there is space at the end of your life to wonder just what YOU are bequeathing to the next generation — what have you added to the world that will last? And what have you taken, too? How much forest or land destroyed for the things you have consumed that the next generations will never know?*

*How may trees have you planted? What dreams and love have you inspired? What ideas of yours will influence children to come and what buildings do you bequeath to the next generation? Most old buildings are precious not because they are old, but because they are beautiful, built in the days when people loved what they did.*

*The only reason anyone will preserve a mass-produced modern house is for its museum value: "Good grief, did they really live in things like that?" They won't last because they are loved or are beautiful. When you build a house of stone you are building something special, something that WILL last. I sometimes dream about the generations that will live in this house after I'm gone, and wonder if they will remember me.*

much less. Another three years of working and building and you should be able to live in it. But of course most people who are working spend most of their money on the things their workmates spend money on: clothes, pubs, meals out, frozen (horrible) pizzas, holiday 'escapes', a car they can probably do without . . .

(I admit I'm biased though . . . new cars, holiday 'escapes', weekends away, meals out always seem to me to be compensations for a life you really don't like. Who was it who said that a compensation is something that doesn't quite compensate? A frugal life doesn't have to be poor and cheerless . . . not if you've got a bit of gumption and three spare brain cells and imagination.)

**We spent a year carving the house site out from the hill, mostly by hand. This shows the first stage completed.**

Having been lured into building our own house by the promise of cheap housing, though, most of us find something more.

I was reading the results of an experiment the other day — about pigeons pecking for their food. Apparently the more they had to work for it the more they appreciated it . . .
Which of course is something that humans have already known for a long time — the more work you put into something, the greater the joy and satisfaction when it's done.

## The beginning

This was never meant to be a house. It's a pretty nice house now (sometimes a lovely one). But in the beginning it was just a steel framed machinery shed/packing shed that we built to park the tractor in and service what we hoped would be productive orchards. The 'real house' was going to be on the other side of the creek.

David and I had been dreaming about the 'real house' for years. It was going to be made of stone, because stone is solid and lasts for centuries and there was a heck of a lot of stone on the farm, which meant (we hoped) we could build the house for practically nothing, because after buying the land (and working two jobs at a time — three if you counted trying to farm) nothing was about all we had left.

### DELAYED GRATIFICATION

*This is a 'want it in an instant' era. TV tempts us to bright and beautiful things (or they look that way in the ads anyway) and trains us to think that all we have to do is fork out cash and get it. I don't want to sound like a fanatic here, but I think our culture is forgetting that hard work is actually enormously satisfying. I find the greatest pleasure in life is to work long hours at what I love doing, and have achieved something at the end of it.*

*The more you put into a project, the greater the thrill. Bryan always knows when I've finished writing a book because I wander down to his shed going: "Oh wow, oh wow, oh wow!" and he says, "Ah, you must have finished the first draft . . . "*

*Of course, there is work and there is work. Many jobs today aren't really producing anything necessary, there isn't a buzz or sense of achievement at the end of the day. My cousin, who works in finance, once said: "At the end of the week all I've really done is contributed to Australia's gross national debt".*

*Building a stone house IS work: heavy, dirty, sweaty and the cement dust gets up your nose and eats little holes in your soft flesh (use lots of hand cream: I found comfrey ointment magic when I was doing lots of concreting). But the buzz at the end of the day, as you kick the new stone wall and know short of a bazooka it will never fall down . . . is extraordinary.*

### TIME

*Stone walls — once you get the hang of building them — take about the same time as building mudbrick or rammed earth walls and much more time than building with timber. Our walls took about three months altogether to build (a lot longer than that in fact, as they weren't worked at constantly). But again, the walls take only a small part of the time it takes to build a house. The foundations and footings take by far the longest amount of time, followed by plumbing, carpentry, indoors, and things like tiling.*

So we spent a year carving a house site out from the hill, mostly by hand (this valley is pretty much vertical — the only more or less flat land was too valuable to use as a house site). The foundations were dug, the plumbing glued together, the septic tank hole dug. This all took another six months while we saved up to buy the reinforcing and buy the concrete for the slab. The concrete slab was going to be our one extravagance — genuine, professionally poured concrete, so the entire slab would be poured in one day. A slab that is poured in several goes is never quite as strong — or as termite proof — as one that is all poured at once.

We wanted our house to be strong.

A host of friends turned up on 'pouring day' — Richard and Val (we'd helped pour their slab six months before) and the Smiths and . . . it's too far back to remember exactly who was there now. The list also included a sweet chap who we'd turned in to the police the year before for shooting 'roos on our place. (Well, he didn't exactly shoot any. He missed.)

We hadn't ever seen him again but he turned up cheerfully anyway, saying he'd heard we might need a hand and anyway he'd reformed. He didn't shoot 'roos for dog tucker any more, only sheep, they were the real feral pest in Australia . . .

So we waited for the concrete, while the stew that was to feed us afterwards bubbled on the stove in the tin shed where we were living till the house was built, along with the cake and slices and other stuff for hungry concrete layers . . .

And we kept waiting.

The concrete finally arrived, about two hours late. The first lot anyway. We'd specified that it was to be carried dry and mixed on site — concrete churned too long loses its strength. But it was wet and brown and didn't really look like concrete, more like sort of muddy sand.

The next lot arrived . . . and then the truck broke down. It turned out it didn't matter that the rest of the slab wasn't poured that day, because the concrete was no good anyway. In fact the expert who came to test it for us said it was the worst lot he'd ever tested, with the compressive strength of wet sand. (We'd got suspicious when it crumbled away when we jumped on it.)

There was no way it could ever be used for a stone house — not even a wooden house, unless we were prepared to dig new foundations around the edge of the old slab, and even then we'd need some fancy engineering.

The concrete supplier denied there was anything wrong with it. He sued us and we sued him, and it took years to get to court. He paid us just as we were going in to the courtroom. The sad thing was I don't think he ever really accepted that there was anything wrong with his concrete, despite the tests.

The slab is still crumbling away, deserted among the wattle trees in what was meant to have been the garden.

By the time we got to court I was pregnant, and the marriage had broken up once but was shakily together again, and we'd done the only thing we could afford to do while the case was being settled — build stone walls around the machinery shed across the creek, and moved in there temporarily.

Two decades later I'm still living there.

Now it's a house, not a shed. I doubt that anyone who didn't know its story would guess that it ever was a shed. Many people ask us what year it

### THE COLOURS OF GRANITE

*There is a stone wall about a metre from my left arm, as I sit here tapping away. It sounds really boring, when you put it like that, as though stone was like concrete or bricks, all pretty much the same. The more you look at this stone wall (and I gaze at it pretty often as I hunt for something to say) the more colours and shapes you see in it: pink grey red, a hundred different yellows and ochres, a touch of green, some purple, dark brown and one rock the colour of old blood.*

*Every shape is different too, even though these rocks were selected to be flat faced and mildly rectangular, though none of them really are. I've been staring at this wall, on and off, for 20 years. With luck I'll spend another half century studying it. I'll bet I'll still find new aspects of it even then. How can anyone be inspired next to a wall of pale brick?*

**Stone houses look and feel solid. This is the house that Sue McIlveen and Richard Allen built.**

was built, expecting us to say '1840' or '1862'.

One righteous woman even took us furiously to task during the last lot of alterations, for destroying a part of Australia's architectural heritage.

"Exactly when was this house built!" she demanded.

"1981-2," I said.

She shut up after that.

## Stone walls

I have always loved stone — stone somehow anchors you to the landscape. Great rocks in paddocks, tumbled boulders nestled on hills, beaches of pebbles or stone glinting at you through the water. I collected stones as a child, even though the colours faded out of the water, and the book shelves in my study are still littered with rocks.

For me stone is one of the essential elements of life, like water and wood and soil and fire and animals and trees. I'd hate to be without any of those in my life.

Stone houses feel solid, even if you know that they're really no more secure than brick or wood. You can gaze at a stone wall for hours, learning each peculiarity of every rock — a lovely thing to do on wet winter days when the firelight flickers across the walls. You can kick a stone wall and know you won't damage it at all.

There are three more commonsense advantages to building out of stone.

Firstly, it's solid.

Secondly, it's cheap, as long as you have access to the stone. Our creek is full of tumbled boulders and this end of the valley is full of mullock heaps too — great piles of rock turned over by the gold miners last century, sometimes left in giant piles, sometimes carved into walls or neat stone trenches or the odd strange structures whose function is now inexplicable and forgotten.

Thirdly, you either build a stone wall properly or it falls down. This is a great advantage for a novice builder. If your work stays up you've done it right.

David had grown up with stone, watching Rocky Campbell build freestone walls on his family farm in Tasmania. The only rocks I'd ever touched were ones I'd tried to skim across the waves. Neither of us had ever built anything out of stone.

But we had read *Living The Good Life* by Helen and Scott Nearing, old-style communist idealists who had chosen back-to-the-land self sufficiency in the 1930s and were dedicated stone builders. (My friend Val says that their book should have been titled *Living The Good Life With 1,000 Tonnes Of Concrete.*)

I liked the Nearings' conviction that the good life involves physical work as well as mental work; that humans have bodies as well as minds and need

### Roof on or roof off

*This house is a converted steel frame machinery shed, three wide bays, a roof and a dirt floor. We designed the machinery shed ourselves, got council permission, then arranged for an engineering works found through the Yellow Pages to build it for us, in chunks that we could carry in the green truck then put together ourselves, a bit like a giant version of Lego. This meant that the roof was already on when we began to build the walls.*

*If I were building again I'd probably put the roof on last — building that last metre of stone work under a roof was no joke, trying to wriggle rocks under the tin. My neck felt like it would never be straight again. But at least we could store bags of cement, doors, windows and other building material under shelter while we worked, not to mention sheltering ourselves when it rained. And as the walls rose the place really did start to look more like a house than it would have if it had been topless.*

*Come to think of it, maybe I'd still put the roof on first. I dunno . . . I do know, though, that I'd still get the roof framework done professionally, whether it be wood or metal. It doesn't add much to the cost, and it means you don't have to be leaping up above the ground like a mountain goat.*

*But basically if you can follow a plan you can put up a roof. Many friends have done it, with no previous building experience. I just don't happen to be one of them.*

### Kids and building sites

*Building sites are dangerous, especially for kids. On the other hand, teaching kids how to build gives kids confidence plus excellent skills for later in life. It's also fun. One of today's tragedies is how much kids are separated from the two basic life skills: how to build their own home and how to grow their own food. Learning these teaches kids that work can be fun and enormously productive, rather than just meaningless slog for money.*

*Do remember though that if kids are on a building site they must be closely supervised and taught, not just left to play batman on the scaffolding or mud pies with the bags of cement. Most tools are not kid size; most scaffolding isn't kid size either, and this makes them much more dangerous for kids than adults, especially as kids have less experience.*

to use both of them productively; that all parts of your life can be a creative challenge, and that we owe a duty to the next generations to build well and solidly. (Some of their other convictions — including living on mostly whole grains and raw vegies — I've never really taken to.)

## Formwork walls

The Nearings had adapted an earlier technique for building cheap, fast, solid houses, using formwork to keep the walls straight. Conventional stonework needs skill, and time to set each stone exactly as it should be. Formwork stone is so easy it's almost impossible to get wrong.

This was the method we used to build all the walls. It took a little practice to get an attractive result — the last walls we built were much better looking than the first. But it was simple, reasonably fast (not as fast as wooden or Colourbond walls that you just nail up, but probably as fast as mudbrick or pisé or strawbale walls).

Once you get your mind around it all, it's really remarkably easy to do. (I have possibly less mechanical aptitude than anyone else in Australia — but I love building formwork stone walls.)

## The perfect block of land

The perfect block of land is the one you fall in love with . . . but make sure it is true love!

If you've never had a place of your own before you may THINK you have fallen in love with the first bit of land with two gum trees on it that you see but, like your partner in life, the first one you fancy probably isn't going to be the last.

Whether your land is in the city or country, if possible look for:

1. A gentle slope that faces north.

2. Water SOMEWHERE near . . . humans feel happier with water nearby, as long as it isn't going to flood them. If necessary add your own, out the front where you can see it as you enter — a dam or fountain depending on circumstances.

3. Check danger of flooding. Theoretically your local council should prohibit building on flood prone land. In practice this doesn't happen. Check where past floods have come to. As areas are developed with more concrete, less grass and more run off, flooding can also drastically increase. You may

## Where to put a stone house

*Stone houses are heavy and not particularly flexible. They are best built on solid ground — not ground that has been recently filled from excavation. If in any doubt at all, get an engineer to certify that your site is suitable.*

*Also make sure that your soil is well drained and stable. Avoid stone houses in earthquake prone areas, unless you have a very, very good engineer who says your design is safe.*

even find that on a presumably safe slope run off from houses above you can form a wall of water that rushes through your place.

4. Make sure you're not overshadowed by something subliminally threatening — a taller house, flats etc. Also make sure there isn't too much likelihood of that occurring!

5. Fire usually runs faster uphill than downhill; houses on slopes may be more at risk, and some areas are far more prone to bush fire than others. A stone house will not make you invulnerable to fire. Do check the fire history of your area; ask you local fire authority for their suggested prevention measures.

6. Look around at the potential for development around you, especially traffic noise. Can your road become a major arterial road? Will flats mushroom up around you? Don't fall in love with a view that may turn into 200 brick boxes with letterboxes and driveways.

7. Don't worry too much about steep slopes or any problem that can be countered by human ingenuity! They'll just make the block more affordable — and with a bit if creativity you can overcome them.

### Preparing the house site

We levelled most of our site by hand, using mattocks and wheelbarrow and shovels and crowbarring out the rocks. Towards the end of the proc-

## How much sand, gravel and cement will you need?

*For a slab and footings, for each cubic metre of concrete you'll need about eight bags of cement, half a cubic metre of sand and about one cubic metre of gravel. If you remember your school maths you should be able to work out how many cubic metres are in your slab, and how many in your footings.*

*Simply multiply the length by the depth by the width of your slab (or footings) to arrive at the number of cubic metres. Remember to check that all units are standard (that is do not confuse the millimetres of the depth with the metres of the length – convert all dimensions to metres before doing your multiplication).*

*For complex-shaped slabs divide them up into simple squares and rectangles first, multiply their three dimensions (length, width and height) together and then add up the cubic metres of all the small pieces to find out what the total requirements will be.*

*Unless you are an accountant or statistician or someone really comfortable with numbers, it is really useful to have two people doing this independently for a large project and then comparing their totals! This way you can keep your errors within acceptable limits rather than have them blow out by a factor of ten or a hundred because of a simple arithmetic or logical error that has compounded.*

*Large mistakes, which lead to massive over-ordering, can be very costly. Mistakes in the other direction can cost you time and friendships (when you have assembled the willing workforce for a day of messing about with concrete and run out of the necessaries by 10 am). Not everyone is prepared to drive for hours a second time to 'help' when they hung about with nothing to do previously.*

*For stone walls the more stones you fill your wall with, the less concrete you'll need and it's very hard to work this out accurately in advance. It is worth doing a rough estimate though, so you get some idea of the eventual cost. As a VERY rough rule of thumb about 10 per cent of your wall will be concrete, the rest stone . . . but hopefully it will be much less than that!*

*Don't buy all your cement in one go. Buy only what you will use in a fortnight or so, and DO keep it well covered and on a waterproof mat of plastic. Otherwise you may find your expensive concrete has turned to rock before you get it in the house.*

*Do check all bags of cement before you take them home too — a surprising number may have hard corners where they've become damp. There is no point paying for stuff you can't use.*

ess we bought an old secondhand tractor with a bucket — which promptly broke as it strained under the weight of yet another granite boulder.

You can level sites by hand or hire a bulldozer. If your house site is on a slope, you'll find that a lot of dirt has to be dug out. DON'T build stone walls on this. Push the dirt further out to make a gently sloping garden as it subsides, or to each side. Stone houses shouldn't be built on 'back fill' unless it's had several years to compact. They are best built on level ground (or at least several levels of level ground if you want a 'stepped' house). Beware of land that was previously mined or if there are caves in the area or any other soil disturbance.

There is no such thing as a perfectly level flat site! In fact a slight slope, say 1:50, helps drainage.

Use a spirit level to make sure your site is level. We placed blocks of wood around the site, stretched a wooden beam from one to another and laid the spirit level on the wood. I suspect though there are more professional ways of doing this!

If you do keep some slope, you will find that one part of your formwork for the footings and slab is higher than another, so that your slab will be thicker in some parts than others. This is fine — as I said, a slight slope gives better drainage — but the more slope there is, the more concrete you will need, and the more expensive your slab will be.

If you plan to grow fruit trees etc, do plant them before you start your house — the sooner they're in, the sooner you're munching fruit. But don't — repeat don't — try to establish any garden that needs a lot of looking after till you've finished the house. And definitely don't plant anything where some muggins will tip a barrowload of concretey water on it, or a delivery truck will drive over it, or someone can tip a tonne of gravel on it.

## MAKING YOUR STONE WORK LOOK GOOD

### COURSES

*When you are building with 'dressed stone', that is, stone carved into large brick-like shapes, you lay 'courses': one layer, then another, then another. When you are building with forms you don't have to do that . . . you can just bung them in!*

*Stone walls do look better, however, if you try to stick to 'courses' — lay one layer of rock, bed it firmly, then do another, taking great care to try to 'tie' the walls together, that is, place a rock over the gap between two rocks below.*

*You don't actually NEED to do this for a stable wall using formwork but if you don't it won't LOOK stable when you've finished.*

### BEDDING

*When you see a rock in the paddock it's sitting as though it grew there — firm and secure and 'bedded'. If you sit that rock up on its haunches or in some unnatural way, it won't look good. ALWAYS place rocks the way they would naturally lie on the ground, in their original plane, not up on their sides or tilted dramatically. Stone laid properly gives an incredible feeling of safety and solidity. Stone laid incorrectly just feels . . . odd.*

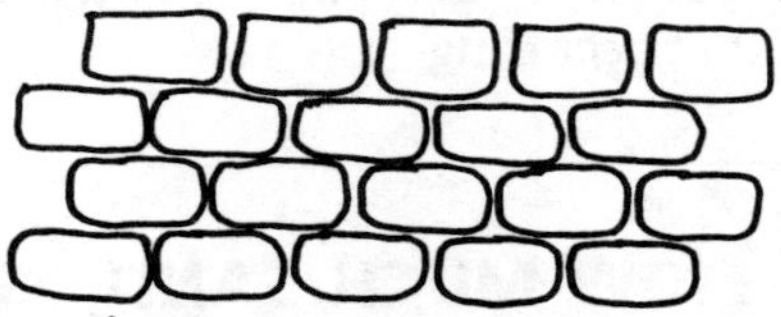

laid squared off stone with mortar.
no formwork needed.

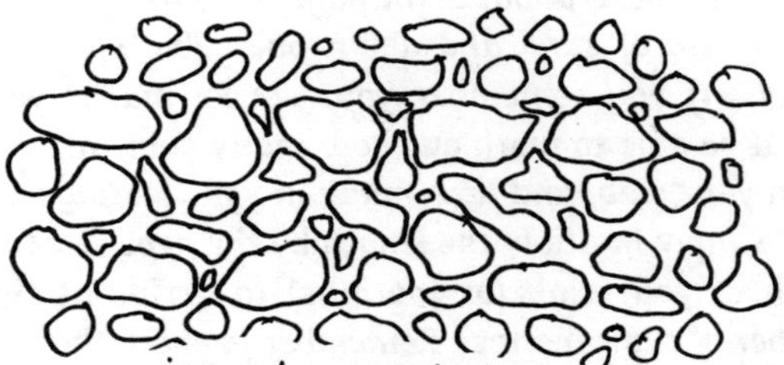

irregular paddock rock laid in formwork without any attempt at courses

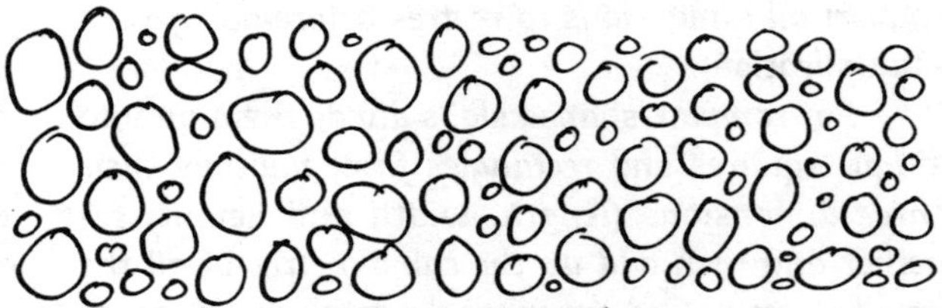

'random rubble' rounded river rocks in formwork without any attempt to lay them in courses

narrow split rocks in formwork

## COLOURING CONCRETE

*When it dries, concrete ranges from grey to white. Some people like the contrast between dark rock and paler concrete; others think it makes the house look like it has the measles.*

*It is very simple to change the colour of concrete. You go to the hardware store, buy a small bag of colouring, and add some to each batch of concrete you mix, making sure that it is the same amount each time so that each batch matches the others. Do be discreet too; you will need a surprisingly small amount of colour material.*

*You can buy yellow, blue, green, red, brown and black colouring materials. If you mix these together you'll get other shades too: a rich ochre from brown and yellow, orange with yellow and red, dark ochre with red and black.*

*We didn't add colouring to the concrete in our stone walls. I wouldn't if I were building them again either but it's good to know the option is there if you want it.*

*We did add a dark brown red colouring for the concrete around the paving in one of the bedrooms, and it does look good. I'd also add colouring to:*

- *concrete paths*
- *homemade pavers*
- *patios*
- *visible bits of your foundations — a touch of pale yellow ochre will make them less conspicuous; a touch of bright blue might make them a feature!*
- *smooth concrete floors such as in a laundry. If you have a rich blue concrete, or a dark ochre concrete, you might decide that you don't need to pave, tile or cover it with vinyl or cork.*

Building sites are ALWAYS messy. It's best to go with the flow and make them attractive when you've finished.

### Slabs

Stone houses can be built either on footings — trenches where the walls will be — or on footings plus a concrete slab between them.

I do like slabs. They give you a solid foundation for flooring, retain heat or stay cool wonderfully, have fewer problems with termites . . . but that doesn't mean that there are NO problems. Do investigate Termimesh or one of the other termite defence systems BEFORE you build. I can't think of any area in Australia that doesn't have termites.

Slabs must be raised slightly above the ground to help run off in storms. This raised profile also makes it easier to check for termite tunnels.

The details of your slab will depend on the size and shape of your house, where it is set and local building regulations, plus possibly the advice of an engineer. But for general slab curing techniques, see 'Laying rock paving in concrete' on page 59.

### Footings

You can hire a backhoe to dig your footings, or do it by hand. I'd recommend the latter — even if you hire some of the hands. It's less mess but I'm biased against machinery, so take my advice with a grain of salt.

Again, if I were starting out I'd try to get the money together to pour all footings and foundations in one day — as we tried to do with the original house we tried to build.

When we came to do the footings and foundation for this place, all our money was tied up in the crumbling concrete on the other side of the creek. The footings and foundations were dug out, the plastic and reinforcing laid and then we hand-mixed and laid the concrete bit by bit, leaving I'm sure structural weaknesses and places where termites, damp etc can penetrate (though as far as I know they haven't yet).

If possible, do it all in one hit!

Stone walls need reinforced concrete footings. (Yes, I know that many ancient stone buildings don't have these. A heck of a lot of footings-less stone buildings ceased to exist before they became ancient too.)

Footings need to be 10 cm wider on each side than the width of your wall so that you can balance the formwork on them. If your wall is going to be 60 cm wide, then your footings have to be 80 cm wide. Our footings are 50 cm wide (for 30 cm wide walls) and 100 cm deep.

The depth of YOUR footings and foundations will depend on the height and thickness of your walls and your local building code. They will also depend on the type of soil you are building on — clay soils, for example, will need deeper foundations than ones on loam, as clay swells and shrinks. Consult an engineer, and consult your local build-

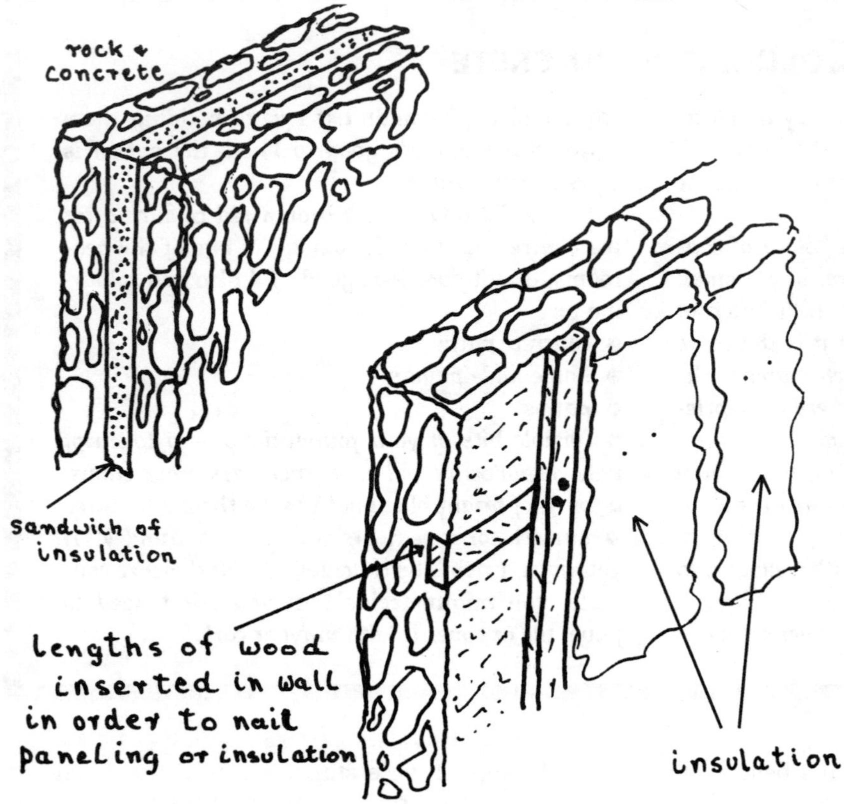


**EXTRA INSULATION FOR STONE WALLS**

ing inspector — in most cases you'll find them incredibly helpful (you may, of course, find the odd one going through a mid-life crisis but this is rare). Building inspectors are also much happier if you let them know exactly what you are planning and how it is progressing if you are building something unconventional. They don't like to be taken by surprise when a councillor demands — "What are those hippies out on the Back Blocks Road up to?"

In most cases they'll give you an enormous amount of extremely helpful advice, including the best way through a host of rules and regulations. They will also tell you how large your footings have to be, how much reinforcing you need to use etc or at least how you can find out.

Don't be afraid to ask (also don't hesitate to at least write a letter of thanks, not to mention the odd bunch of flowers or apple tea cake).

## How thick will your walls be?

Stone walls need to be at least 30 cm thick to be stable. The thicker your walls are, the cooler you'll be in summer and the longer it will take for heat to escape in winter. You also get wonderfully deep window casements! Very wide stone walls on the other hand mean more work, more stones and more concrete!

Our walls are 30 cm thick. I think I'd make them a little thicker if I were doing it again, mostly because I do love really deep window casements and doorways, and the way the light pours through them and the shadows move across the floor.

It takes about three stinking hot days plus hot nights for our walls to feel warm to the touch (has only happened three times since the house was built) and about three cold cloudy days without the fire on inside for them to cool right down. In terms of cooling and heating in our climate, I don't think the walls need to be thicker — but they may need to be thicker for your climate.

## Insulation

Stone may seem like a fantastic insulator. It isn't.

Stone is a great heat sink — or cool sink — but if you want a really well insulated house, consider placing a thickness of insulation in the middle of the stone wall, or embedding strips of wood on the inner side so that insulation and wood panelling to cover it can be nailed on later.

If you don't want to do that, make sure that no stone extends more than two thirds of the way through the wall. Concrete is a better insulator than stone — but even concrete isn't particularly good.

We don't have insulation in our walls, except the wooden ones, and if I were building again I don't think I'd bother. Okay, neither stone nor concrete are good insulators — but a thick stone wall does have a great deal of mass and thermal inertia, so there is always a considerable time lag between the day warming up outside and the heat penetrating through the walls, or the warm air indoors escaping out to the winter air.

If I lived in an area that was hot or cold day and night, I'd insulate my walls. But as our summer nights are cool, and our winter days usually warmish, we just open up the house or close it down with shut windows and pulled blinds as needed. However, we do have as much insulation as possible in the roof. Don't skimp on it.

# TWO

# The Formwork Stone Wall Method

## Step 1 — Preparation

**Find the rocks.**
**This is the most time consuming part of all.**

### Which stone to use

A rose is a rose is a rose . . . but stone isn't necessarily rock solid. Avoid any rocks that crumble (they do exist) or split easily. Some stone is relatively soft and crumbly when it's dug out, but becomes harder later. Bash your rock with a sledgehammer before you decide to use it. If it crumbles, shatters or goes phut, don't use it.

Sandstones and limestone can often be easily split to make good straight surfaces and even corners. On the other hand limestone can weather badly in very wet areas and should be protected by wide eaves and verandahs.

We built our house from granite. Granite is solid and stays solid. In other words, forget about shaping the stuff, unless it already has a useful crack to split it!

### Where to get rock

Note: do try to use just one sort of rock in your house, or if you must mix different rocks, make the mix into a deliberate pattern particularly around doors and windows.

#### Creeks

Creek rocks are usually rounded — not a terribly good look in walls (though some people love the rounded pebble look) but at least they are clean, and with much searching you will find rocks that have split with straightish sides. Avoid waterlogged rocks or at least give them a few months to dry out. Don't take too many from any site though — you'll be disturbing the habitat of too many small creatures, and may also affect the flow or path of the creek. Also — make sure it is legal to take rocks from a creek bed in your area — it may not be!

#### Paddock rocks

Paddock rocks provide valuable habitat — you may well be making a small army of creatures homeless while building your own house. If, however, you know that the rocks will be removed by a farmer anyway, they may as well end up in your house.

Paddock rocks are usually attractively weathered, often lichen covered, which must be scrubbed off and will come in many shapes and sizes. You'll end up with a rustic, varied looking stone wall. (You can, however, leave the lichen on the outside face of the rocks, for an instant 'aged' look.)

#### Rocks by the Road

There are a surprisingly large number of rocks along roads turned up by roadworks, both little 'fillers' and larger ones. I'm not sure it's legal to collect them, so of course I'm not advocating it here . . . but a recently upturned rock won't yet be a major part of the ecology, and a few hours foraging can net you a surprising amount. (I am of course speaking totally theoretically.)

#### Quarried Bluestone

Bluestone is quarried from basalt deposits. It's a dark blue grey to black — like roads. You can sometimes get excellent reject chunks cheaply if you live near a quarry.

#### Quarried Sandstone

This will come in lovely great giant bricks, easy to use and fit together with nice neat corners. Like bluestone, quarries often sell 'seconds' or 'rejects' quite cheaply. These rejects may not be good enough to make a freestone wall, but they will certainly be suitable for the formwork method.

#### Old House Sites

DON'T loot old stone house sites for your rock, unless you are sure it's all about to be bulldozed away. These sites may be valuable historically, sentimentally or for later archeological studies . . . leave them be!

## Preparing your rocks

First of all the rocks must be CLEAN. If they are dirty the concrete won't stick them together properly.

If they have lichen on them, the lichen will keep 'breathing' even while the walls are in the wall, and in a couple of decades your stones will pop out of the walls. (Val Plumwood carefully kept the lichen on the OUTSIDE faces of the rocks she used in her house; the house looked wonderfully aged straight after it was built.)

**Creek rocks are usually rounded but at least they are clean, and with much searching you will find rocks that have split with straightish sides.**

I spent two days and three scrubbing brushes scrubbing lichen off rocks from a scree slope behind the house site, then decided that we'd use nice clean rocks from the creek.

Creek rocks aren't as good as paddock rocks — they may have a high moisture content from long soaking and, even worse, are more rounded than flat and round rocks are hard to work with.

You can still use round rocks with the formwork method, but without at least one flat side you get a wall that looks more concrete than rock.

Stone walls made of rounded rocks, even if they have flat sides, never look quite stable . . . more

### Rock hunting . . . the joy of dust and blackened fingernails

*There is the rock that Libby gave me for my 21st birthday. She carted it I don't know how far, and left it on the back of the truck as it sat in the car park as a surprise, with a bow around it. She didn't worry that someone might steal it before I got to it. Who would bother pinching a giant rock, with or without blue bow? It's a lovely rock, a sort of cream and peach, flat on two sides with a 90 degree edge . . . a perfect corner rock.*

*There's the rock that . . . what was his name? He fell in love with it as we clambered up the gorge one day. It was gleaming green and red at the bottom of a pool, so he swam down to get it and lugged it back over two hours of rock hoping, clutching it to his naked chest. He almost gave up half a km from home, but persisted. The colours faded a little when it dried but how can you not love a rock like that?*

*Hunting rocks grabs you, much like big game hunting I suppose. There's a triumph every time you find a perfect one. The addiction grows so you make the most ludicrous efforts to bring home a rock you fall in love with, sneaking up with crowbar and gloves to lever out a perfect roadside rock by torchlight.*

*Rocks are such solid things. And the more you learn to really look at them, much like you'd look at paintings in an art gallery, the more you prize each one. To be honest I'd prefer to look at rocks now than paintings. I've learnt to appreciate the subtleties of rock now, but I'm just an amateur at understanding art.*

*Hunting rocks is dusty, frequently sweaty, leads to blackened fingernails and crook backs unless you're careful (I wasn't). So what? Art galleries lead to sore feet and sometimes Legionnaire's disease, but they're still worthwhile. A rock is . . . blast it, there are no real words for it. When you've built out of stone you'll understand. If you haven't, nothing I can say will be enough.*

*Oh, and watch out for funnel web spiders and snakes.*

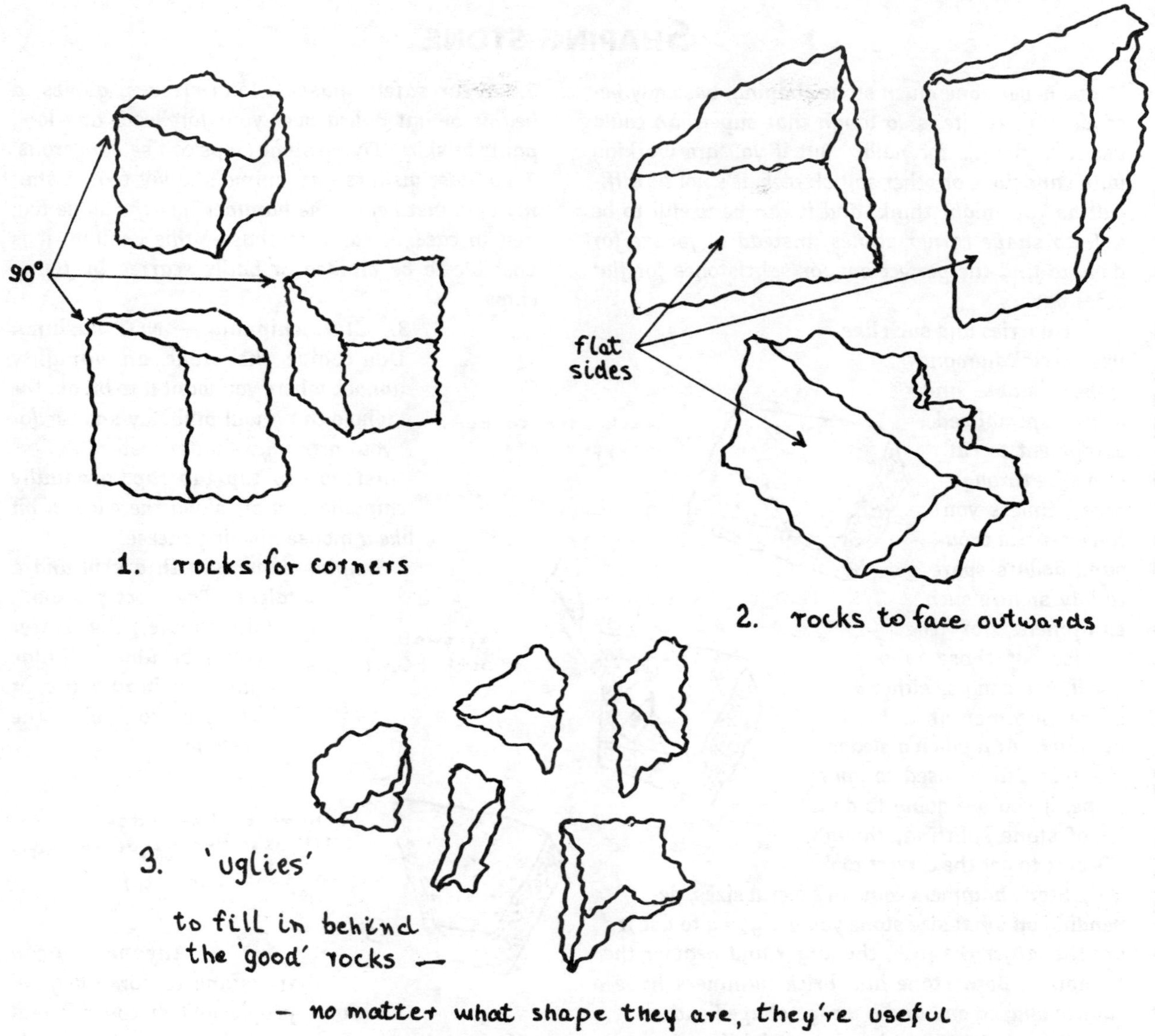


**MAKE THREE PILES OF ROCKS**

like a wall with the measles than old-fashioned stone work. If possible, do try to get roughly rectangular rocks.

You need a lot of rocks for a stone house:

- rocks with one flat side for pressing up against the formwork — these are the 'public' rocks that will form the outside of the wall.
- rocks with at least one brick-like, right angled corner, for the corners of the house. You can align two flat rocks together at the corner, but it looks messy. Neat corners add a lot to the final impression.
- 'ugly' rocks — large, medium and small, for pushing between the other rocks so you use much less concrete.

We made three piles of rocks — one pile for each category, with another pile of rocks that might make good flagstones or steps for later use, plus one rock that had a nice sized gold nugget in it, but which got bunged into a wall by accident.

## What size rocks?

The larger your rocks, the less time it takes to build . . . and the less concrete material you have to buy. On the other hand, the larger the rock the sorer your back will be, and they can be difficult to manouvre safely.

The size of the rocks you use will also be governed by how you want your house to look. Do you want a nice even appearance with rocks of much the same size? Or a mix of small and large ones? Do you like the idea of a few feature 'giants' in your wall?

## SHAPING STONE

*I have never done much stone shaping, basically because our granite is so tough that superman could use it to file his toe nails. But if you are working with sandstone or other softish rock, it's not as difficult as you might think, and it can be useful to be able to shape corner stones, instead of forage for days to find the perfect one, or split stones for flat sides pavers.*

*Quarries and such like use special diamond-tipped saws and other specialised equipment to cut and slice through rock. Unless you have several thousand dollars spare to buy or hire such equipment, or the services of those who use it, you can use either a stone hammer or a brick hammer. At a pinch a sledge hammer can be used to split stone; if you are going to do a lot of stone splitting, though, it's best to get the correct tool.*

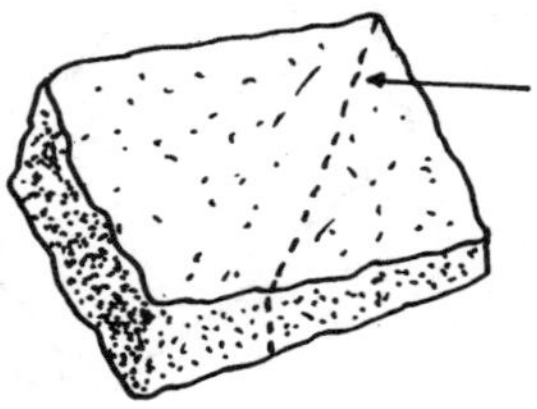


1.

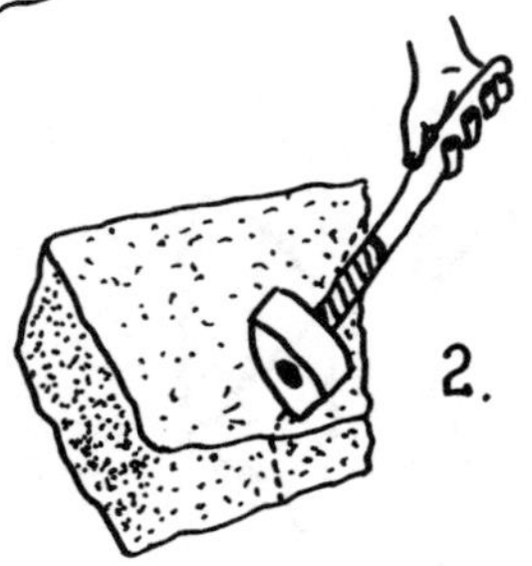


2.

*Stone hammers come in several sizes, depending on what size stone you are going to use it on: the larger the rock, the larger and heavier the hammer. Both stone and brick hammers have a square-shaped end and a wedge-shaped end.*

*The wedge-shaped end is useful if you want to split a rock along a useful crack, or, if you are particularly ambitious and dexterous, to split stone along its grain, much like you'd split wood along its grain, to get flat slabs for paving or facing walls.*

*Most stone shaping, though, is done with the square-shaped end.*

*1. Mark where you want to chip along to.*

*2. Wear safety glasses, long sleeves, gloves, a beanie or hat pulled over your forehead, and long pants or skirt. Flying stone chips can be dangerous. Keep kids, visitors and animals away too. I shut my eyes just before the hammer hits the stone too, just in case. I can't emphasise this enough: it is possible to be blinded or badly scarred by flying chips.*

*3. Start chipping — NOT bashing. Don't give the stone an almighty whack where you want it to break: the whole thing will probably shatter (or your arm may shatter instead). Just go tap tap tap tap, gradually chipping bits off along the edge, a bit like a mouse nibbling cheese.*

*Shaping stone is both a skill and a talent. The more you work with stone, the better you'll be able to judge just how hard to tap, or where to split the stone.*

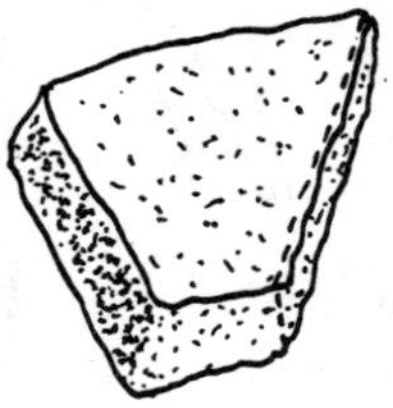


3.

*Anyone can shape stone to some degree. Some people will discover a real affinity with the rock: a good stonemason can do extraordinary things with a piece of stone.*

*Note: wrap ducting tape around and around the handle of your stone or brick hammer. This will protect it somewhat if you hit the rock with the handle instead of the hammer end; it will also help support it slightly so it's less likely to split, and protect your hand from splinters too.*

Tastes vary. I like a more random look in my stone walls, and I do love a few giants with their variations in colour and texture; other people prefer stones all much the same size. But do think about this before you get too far down the planning road.

Once a wall is up, it's difficult to change!

## STEP 2 — GET THE OTHER STUFF!

You will need:

- reinforcing mesh and rods — this can be bought or sometimes scavenged. We were lucky enough to gather stuff thrown away after a giant sewerage tunnel was built — great thick lengths of steel.

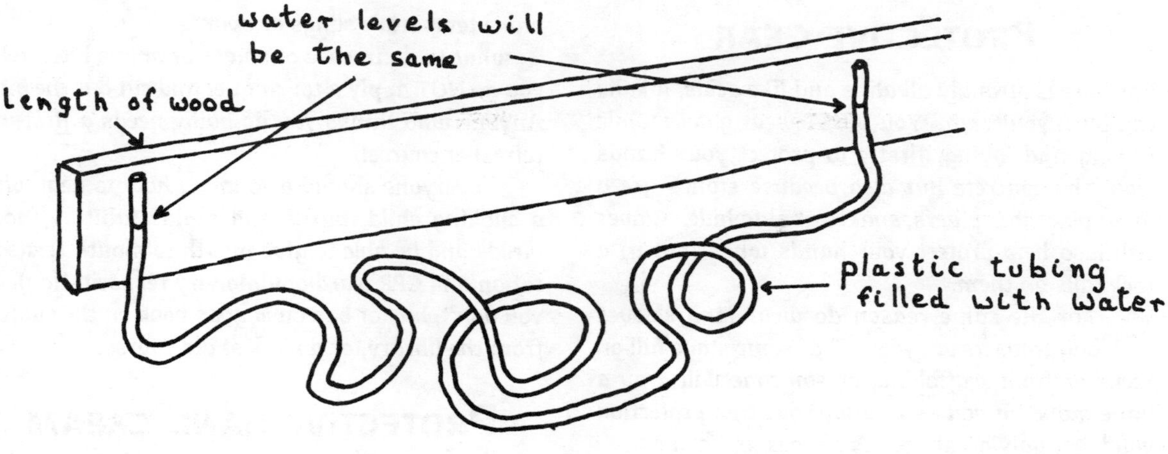


**HOMEMADE LEVEL**

- a wheelbarrow — not quite essential but useful — unless you are a woman in which case it is essential — you can push much heavier weights than you can safely stagger with.

If possible get a high-sided one — a builder's barrow, not a shallow-sided gardener's barrow. The modern black plastic variety are a worthwhile investment and are lightweight, strong, easy to clean and, of course, they don't rust.

DO clean out your wheelbarrow every time you take a break or stop for the day. It is very easy to gradually accumulate concrete in the crevices, and this makes it heavier to use and of course the barrow won't hold as much.

Do dry your barrow after you wash it when work is finished for the day.

Do either turn it upside down after use so it doesn't fill up with water if it rains, or dew condense on it. Keep your barrow under cover as much as possible, out of the sun and rain. Don't just leave it around the back of the house till you need it again.

Don't leave a metal barrow full of damp material like leaves.

(If you and your partner both intend to do a lot of work outside at your place, a wheelbarrow each saves a lot of arguments.)

- many buckets, either some metal and some plastic ones, or all plastic. I prefer all plastic buckets — heavy duty plastic ones (those white catering-style buckets are very sturdy as are the ones designed for washing nappies) for the carting of wet concrete and the light, cheap ones for the water throwing.

The metal or thick plastic ones will also be used for carting concrete to the walls to tip into the forms. Flimsy plastic ones can be used to measure water to throw into the cement mixer. They are much cheaper than sturdy buckets, but in the long run sturdy buckets are more cost effective.

Buckets can also be made out of empty metal drums, with the lid off and wire threaded through the top. Do make sure that there are no sharp dangerous edges, or worn rusty bits that might break or injure someone.

- a source of cleanish water, preferably with a hose. You'll need the pressure a hose can give you to clean out the cement mixer, buckets and other tools at the end of the day, but in a pinch buckets of water from a creek will do (just don't pollute the creek — only dip in clean buckets!).
- one wooden and two pointed metal trowels for smoothing and finishing.
- a hammer.
- a stone hammer for shaping rocks (definitely optional).
- long nails (the metal sort, not fingernails).
- pliers.
- one builders' level, or a homemade level (see diagram above).
- LOTS of gloves — I found thick leather workman's gloves too cumbersome, especially as soon as they gained a layer of concrete on them. Gardening gloves were thinner and more flexible, but did wear out sooner.

Always have several pairs of gloves on hand — you'll wear them out pretty quickly, but it's better to wear through gloves than skin.

- hand cream. This is ESSENTIAL (see next page).
- materials for concrete (see below).
- materials for forms (see below).
- tamping sticks . . . we used tomato stakes. Lengths of reinforcing rod are easier, have no splinters and they are easy to clean at the end of the day.
- sheets of plastic or old tarps to cover machinery

## PROTECTIVE GEAR

*Concrete is strongly alkaline and like acids, alkalis can burn your skin. You MUST wear gloves while mixing and laying, firstly to protect your hands from the concrete but also because stones are a great place for spiders, snakes et al to hide. Gloves will also help protect your hands when (not if) a rock falls on them.*

*For the same reason do wear long sleeves and long trousers . . . you will at some stage fall on rocks or from scaffolding or someone will drop a large stone on you — and without this protection you'll get ugly great gouges on legs and arms.*

*Do wear thick protective boots and socks, not joggers or gumboots. Every person I know who hasn't done this has had at least one serious flesh-ripping accident.*

*Do wear a giant hat.*

*Do keep a first aid kit on hand with band aids, bandages, adhesive tape, crepe bandages, disinfectant, tweezers, rescue remedy, disposable latex gloves, sterile dressing for severe cuts, sterile eye pads, sterile water if someone gets something in their eye, sterile gauze swabs for cleaning wounds, triangular bandages, lots of cotton wool, and a small pair of scissors that isn't used for anything else.*

*Also learn how to use it all!*

*At the very least anyone who is building their own house (or for that matter anyone!) needs to know:*

- *how to clean a wound and bandage properly (for example, don't stick cotton wool or fluffy stuff onto a wound as it will stick).*
- *how to control bleeding.*
- *how to support an injured arm or hand and make temporary slings or splints.*
- *what to do in case of spider or snake bite. (No, you do NOT apply a tourniquet and cut out the bite. Anyone who thinks you do badly needs a first aid refresher course!)*

*Everyone should also know how to deal with a choking child, burns and scalds fainting, nose bleeds and be able to give mouth to mouth resuscitation and CPR (cardiopulmonary resuscitation). If you don't, buy or borrow a good book on the subject from the library, or do a first aid course.*

## PROTECTIVE HAND CREAM

*Bung this on before and after working with concrete and stone, and maybe a few times during the day too.*

*6 tbsps virgin olive oil*
*3 tbsps beeswax (use a beeswax candle or go to a craft store. DON'T use paraffin wax.)*
*1 large hunk of comfrey root, or lots of finely-chopped fresh or dried leaves.*

*Chop and clean the root, and simmer it in the olive oil for an hour. Strain out the root (or leaves). Add the beeswax to the now green/brown oil. Melt wax, take off the heat and pour into small, clean, wide-mouthed jars. Use liberally.*

*If you can't get comfrey root (you can always go and buy a pot of comfrey from a garden centre) make it up anyway! It won't have the healing power that comfrey gives (comfrey contains allantoin, a cell proliferant — so do several commercial, non plant-based, disinfectant creams, so look at the contents at the back) but it will protect your hands.*

***PS.** Don't apply comfrey to broken skin — that's what the regulations say, anyway.*

and new stone work and to rest bags of cement on.
- useful bits of wood — small bits of ply and 30 cm lengths of 4 x 2 fencing wire.
- wire cutters.
- two brushes — one with wire bristles for scrubbing back green concrete and a stiff scrubbing brush for the end of day barrow and tool clean up. You will go through several brushes, so if you're not near a hardware store it's not a bad idea to buy half a dozen to start with.
- gum boots or steel capped boots that will never be the same again after weeks of mixing, pouring and handling concrete. Do NOT work in bare feet, sneakers or thongs.
- two shovels (only one is essential, but it's easier to have two, or even more if several people, are working on the site. People can get very territorial about their shovel).
- one spade (a shovel is great for shovelling sand and gravel; a spade much better for digging).
- one pick (for digging foundations).
- a crowbar (for levering rocks out of where the foundations will be, or persuading stubborn formwork to shift).
- a wide sieve — this is useful if you collect your own sand and gravel, so you can sift out sticks,

stones and wombat droppings; also any leaves that have blown down since yesterday.

- gaffer tape (never goes astray).
- a truck, trailer or ute for shifting stones, sand, gravel . . . plus the tray also makes a great work bench.

A truck or ute isn't really essential — you may be able to borrow one for rock collection, and after all, stone houses were built long before the internal combustion engine was invented, though I suspect a horse and cart was used to cart rocks instead. But throwing rocks into a truck or trailer or ute is a good way to bash and generally deteriorate a vehicle. And this is okay if it's yours, but a bit hard on the person who's lent it to you.

It is almost impossible to gather a load of stones without scratching paintwork and collecting the odd dent. Whether the ute or trailer is yours or another's find a sheet of plywood to fit the floor for the rock collecting forays as it will prevent a lot of damage to the metal tray. A tarp stretched over as much of the tray as possible will help prevent scratches . . . but some deterioration is inevitable.

Don't be tempted to overload your truck or ute. It's bad for the shock absorbers, bad for the tray, and can make the vehicle unbalanced and less safe to drive. Even with a moderate load you may find that it swings or wobbles unexpectedly . . . be prepared to drive much more slowly and cautiously with a full load. (This may seem like obvious advice but having seen three overloaded vehicles swing off the mountain road above us this year alone, it's advice that needs to be stressed.)

- a cement mixer. This can be electric if you have a source of power (a generator will do) or petrol driven. I mixed most of our cement in a cement mixer, turning it by hand, using a great wheel we found on an old chaff cutter (it was great for bust development).

Later we bought a motor for it. The motor did make it easier but much noisier — I'd still go for hand mixing. (Engines also need more maintenance than humans, and tend to break down with less warning! It can be infuriating if you're all ready to start work and the blasted thing won't go . . . human arms are much less likely to let you down!)

Some people prefer mixing all concrete in a barrow as they are infinitely easier to clean than a mixer — unless you are laying foundations, slabs or are having a working bee with many willing hands and backs the barrow is adequate for one or two people working by themselves laying rocks.

## Caring for your cement mixer

Cement mixers gradually accumulate dried concrete in them and over them. This means they gradually contain less and less mix and they are also heavier and heavier to use.

ALWAYS clean out your machine every time you stop for even half an hour. Make sure it's clean inside and out. Always cover it at night and in the rain — most cement mixers are left outdoors and soon deteriorate. Do wheel it indoors every night or if it rains if you can. Do keep it oiled and well maintained.

DON'T lend it to anyone unless you are SURE they will look after it.

Few people do know how to look after tools nowadays and even a week of neglect is enough to ruin a good mixer. (And no one really cares for someone else's mixer as they do their own).

- a scutching hammer and a geologist's hammer and a pair of goggles — these mean you can remove dags of dried concrete, tap off awkward corners and bulges of rock to make stones fit together more snugly and always protect your eyes before small shards of rock begin to fly.
- food and water. I know these seem self evident, but if you are going to spend long hard days in physical work — which few people nowadays are used to — it's a good idea to organise these in advance, because you'll be bushed at the end of the day.

I found slow simmering stews that I could bung on in the morning great for evening meals, as they didn't need any tending at all when I was tired, plus large amounts of good bread with cheese and varied salad stuff to stick on it for lunch. Also keep plenty of fruit on hand for casual refreshment, and a couple of insulated containers of chilled water with perhaps a bit of lemon juice in it. Do be prepared to eat a heck of a lot more when you're building with stone. You'll need it. If people come to lend a hand, do feed them well.

## Scaffolding

Unless you plan to build stone walls no higher than you can lift (this is actually quite feasible — you can fill in the bit above with windows with wooden columns between) you are going to need scaffolding.

Do NOT assume you can climb a rickety ladder carrying a rock, much less a bucket of concrete. Scaffolding is stable, and there'll be room to place a few rocks on it, good ones for facing outwards and uglies to go behind it, plus a bucket of concrete and a bottle of water (for you, not the walls).

You can buy the scaffolding, then sell it later.

You can borrow it.

You can hire it.

You can build it yourself . . . but either way, you MUST have it.

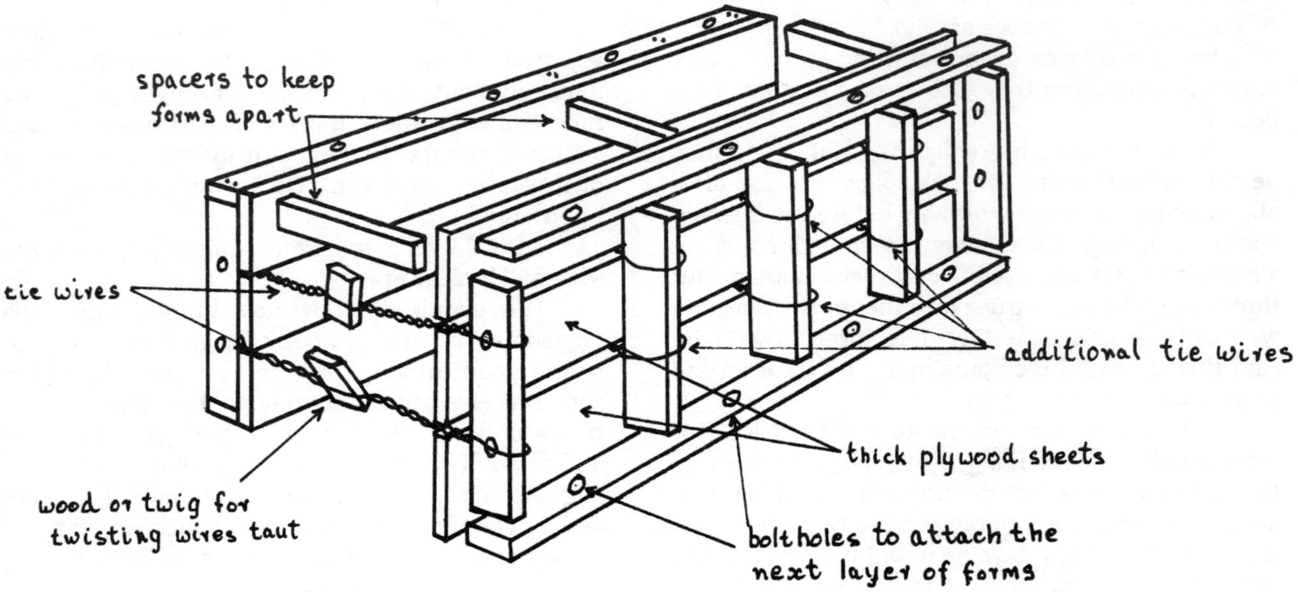


**Basic formwork**

## Step 3 — Build the formwork

We made frames of 4 x 2s (50 x 100 mm timber), about a metre deep and about two metres long, with 13 mm plywood or planks nailed onto the back. (The plywood was cheap 'reject' quality. I've also heard of reject aluminium siding being used to clad forms instead of plywood.)

If I were doing it again I'd make the forms about 450 mm deep and about 2.5 metres long — but be guided by the dimensions of your house. If your walls are going to be 10 metres long, you'll have to be able to multiply the length of the forms to fit into 10 metres, that is, 5 x 2 metre forms. Forms more than 2.5 metres long can be too heavy and unwieldy to move easily.

We used four long forms, and two short ones to fit into awkward spaces. If I were doing it again I think I'd make six or eight forms. The more forms you have, the faster you can work.

If you are going to have the interior wall smooth (just concrete without a rock face), you can make the interior forms much higher. This does save time moving forms around, but a form that is taller than a metre high becomes very difficult to manouvre stones behind — especially if you are short.

Place three or four 'supports' along the formwork to keep it stable. Now oil the forms so that the cement doesn't stick to them and to help stop them rotting. You can use dirty sump oil (and gloves!) which is free, or buy linseed oil for this. I gather diesel also does the job; I haven't tried it — I don't like the thought of working with diesel. I prefer linseed oil, even if it's the most expensive option.

## Step 4 — Scavenge doors and windows

Luckily we were building before the fashion for secondhand building materials really got going. We got some doozy bargains — doors for $3 or just in exchange for hauling them away, solid steel-framed windows. Of course it meant that no door or window matches any other in the house but what the heck?

It's much more difficult nowadays to find extremely cheap scavenged doors and windows. Much secondhand material may even be more expensive than new wood — basically because it's better stuff!

Look in your *Yellow Pages* for secondhand building suppliers; contact kitchen and bathroom renovators who may sell you items they've removed at a reasonable price (don't get your hopes up though) or talk to builders who specialise in renovation.

Also keep your eyes and ears open — if someone is embarking on a major extension or demoli-

**Secondhand building materials can be more solid than new. These steel-framed windows were scavenged from demolition sites.**

tion you might be able to persuade them to allow you to offer your services as a builder's labourer in exchange for scavenging rights.

But be prepared for a couple of hard days' work and to be prompt in removing what you want from the building site.

Relations will sour rapidly if you leave people to work around your pile of treasure/junk while you try to line up transport, storage etc. Nothing is really free.

A friend's sister was given an entire wooden house at the tip one day (she was at the tip — not the house) but the condition was that it had to be removed within a week and she didn't even own a block of land! She bought a block of land the next day and moved the house six days later — this is not the recommended approach!

Also inform everyone you know about your project. Angela managed to get eight matching wooden framed windows because a friend of another friend noticed that their next door neighbour was about to replace their existing wooden windows with aluminium ones and rang them that morning and Angela turned up with a trailer that afternoon.

I once picked up six kitchen sinks at a clearing sale for $6. I only wanted one of them — or perhaps two — but they made good presents for friends who were also building.

If you have time it's also worthwhile chatting with building supply places. Often they have something 'made to order' that didn't meet specifications or materials that are slightly weathered or the last of a batch. You can very often find real bargains. (The floor in my study was the 'end of a pallet' the supplier wanted to get rid of; our front windows were 'made to measure' for someone else but were never picked up. We just changed the proportions of our house plans to fit the windows!).

I'm told untreated wood can rot if it's in direct contact with concrete, especially softwood. The wooden frames in our doors and windows are untreated, and 20 years or so later show no signs of deterioration. But you have been warned.

## GOOD CONCRETE

*When you add water to cement, you start a chemical reaction. If you add sand too, you'll get mortar — a bonding of the cement and the sand. You add gravel to this for strength. The sand/cement mixture will bond to the gravel; and this gravel/sand mixture will bond to the larger stones too.*

*Concrete doesn't 'dry'. It sets, so when the chemical process is completed, it becomes solid. This process takes place more quickly when it's hot but is more efficient when it's cooler. This is why curing concrete should be kept damp for about a week: to slow down the setting process.*

*If the temperature goes below freezing, the water in the mortar will turn to ice. If you are mixing mortar in a cold climate, pour your concrete first thing in the morning after the frost has thawed to give it a chance to set before the temperature drops below freezing again. If in doubt, consult an engineer.*

*The larger the amount of concrete, the longer it takes to set. Stone walls will usually set fairly safely in winter, as long as the ground isn't frozen around them, especially if they are protected by formwork and a trap or other covering. Slabs and footings may be a different matter: do check this out with your local building inspector before you start drawing up timetables! On the other hand there are very few places in Australia where you need to worry about winter temperatures. I just happen to live in one of them!*

*Too much water added to concrete also stops it setting properly. The wetter your concrete, the weaker it will be. Concrete that is too liquid won't support the weight of your rocks; it may also ooze out of the formwork. If your concrete is too dry, however, it won't stick to the rocks or gravel properly and will be hard to spread and tamp.*

*Don't worry about all of this too much. You'll discover what good concrete is as soon as you've mixed and worked with a couple of batches. Just start your wall building in an inconspicuous spot, so that your early mistakes won't be too visible.*

*Our stone walls got better and better as we went along. The first bits are a mess . . . the last bits quite expert. But I wouldn't swap the messy bits for perfection. They are a continual reminder of the achievement of those days.*

### STEP 5 — ORGANISING THE CONCRETE

The formwork method uses concrete, not the mortar that is usually used between stones or bricks: mortar is a mix of six parts sand, one part hydrated lime and one part cement plus water.

The concrete mix we used is one and a half parts cement, three parts sand and six parts gravel. This high proportion of gravel gives a 'stonier' appearance — if you use a sandy mix, or mortar, you can really notice the concrete in between the rocks. With a higher mix of gravel, you don't.

It's important to use as little water as you can to make your concrete strong — and it's difficult to give the exact amount of water you'll need because when you are mixing your own you'll find that both sand and gravel can be damp and so less is needed.

When you first start mixing, add the dry ingredients first — I used shovel loads to measure: six heaps of gravel first (the sharpness of the gravel helps dislodge any concrete remnant from the batch before) then sand then cement, then a measured amount of water in a bucket.

Use about a third of a bucket to begin with then add a little more and a little more as it mixes, till it's a bit like cake mixture or sloppy cow manure, depending on which you are most familiar with. Good concrete should almost stand up in your hand and not slop over it, but it shouldn't be crumbly either. Once you think you have worked out the correct amount of water, make a note for future batches.

After the first batch of concrete has been mixed, add half the water first and whizz the concrete mixer around a few times. This will help dislodge any mixture remaining from the batch before. (There almost certainly will be some unless your mixture has been very wet.) Then add the gravel, then the sand, then the cement, then the rest of the water.

If your batch doesn't mix properly, you may have put too much in. An overloaded machine will have a layer of unmixed sand and gravel at the back. Overloading a machine is also a good way to bust it!

Make concrete just before you use it — don't let it sit any longer than absolutely necessary. Don't let it dry too fast either — the longer it takes to 'cure' the stronger it will be. Cover exposed surfaces with tarps, old blankets, or Hessian potato sacks, and once it's fairly well set, wet it down often.

## Where to get sand and gravel

We used sand from the creek — good coarse stuff, not too silty (never use silty sand). I'd hate to urge people to go out and start causing ecological devastation, though, by digging up sand from the nearest river bed. Our creek has too much sand washing down it nowadays because of overclearing at its headwaters — the more sand that's carted away the better! But do make sure that you're not disturbing useful habitats or going to be arrested.

Don't use sea sand either — too much salt. In a perfect world, you'd make the strongest possible concrete with a mix of fine and coarse sand together.

At one stage we bought gravel by the truckload at the quarry and carted it ourselves. It was much cheaper.

Nowadays we get a large truck to deliver it . . . the larger the load you buy, the cheaper per tonne it will be. Leftover gravel is great on roads or for small concreting jobs (once you start playing with concrete it's addictive — it's so easy and so wonderfully strong and permanent, unless someone takes a bazooka to it. You'll probably find yourself concreting fishponds, fountains, driveways, more rock walls . . . ).

## Step 6 — Dig the foundations

Like most owner-builders, we overbuilt. Our foundation trenches were one metre deep and 50 cm wide. We then laid down plastic, then reinforcing mesh. This time we mixed the concrete by hand. Once I got into the swing of it (and by then I'd already mixed the concrete for a causeway across the creek, plus the foundations for three sheds) I was as fast as the electric motor Bryan and I bought years later.

I'd mix; David and assorted friends would wheel their barrows up and I'd tip it out, then mix another in time for the next comer. It took only one very long day to pour the foundations into the trenches over the reinforcing.

It did take longer to pour the floor slabs. I remember one day we worked right until dark; we were so exhausted that we literally didn't have the energy to float it totally smooth. But without telling us a group of young friends stayed up that night and smoothed it perfectly . . . then left us to discover the magical trowelling the next morning. They were henceforth known as the 'concrete fairies'!

Note: termites can — and do — attack stone houses too . . . unless you are a stone fanatic with a rounded stone roof and there's no wood anywhere for them to eat.

Leave some of the foundations exposed and don't be tempted to cover them too thoroughly with garden. You can then inspect the foundations every few months to check that termites haven't built a tunnel up them to attack the house. It is possible though that they'll attack through a crevice in the slab . . . I once saw a presumed termite proof house where they'd invaded along the inside of a conduit from a generator 100 metres away. Never feel too confident about termites.

# THREE

# Step by Step Stone Walls

WHEN you are building walls with formwork you can either have one side of stone, and the other of smoothish concrete, or both sides of stone. I'd go for one side stone, and the inner side concrete.

Firstly it's faster, and you need fewer 'good' stones. Secondly . . . well, having lived in a stone house for 20 odd years, I know just how much dust, fatty cobwebs and general detritus clings to an uneven wall.

I would now choose to have ALL INTERNAL KITCHEN WALLS SMOOTH so they can be easily brushed, swept or washed. On the other hand, if we had 'pointed' our stone walls and filled in all crevices as soon as we put them up, we could have painted the stone walls with a washable paint. Two of our internal stone walls are now painted, so they can be easily cleaned. The rest have been left as they are.

Internal stone walls are fine for living rooms; they can be a bit spidery for kids' bedrooms unless every small crevice is filled in (and anyway the kids will probably prefer to stick up posters than admire stone work).

But do stick to nice cleanable smoothness in the kitchen!

## Day one

### Step 1

Lay down damp coursing. You don't have to attach it with anything — with several tonnes of stonework on top of it, it's not going to blow away! One twit of an American architect once asked us how we had attached the stone walls to the foundations. We told him we'd left them unattached in case we felt like lifting the house up and moving it one day. I don't think he ever did see the joke.

Balance two pieces of formwork on either side of the foundations. Wire them firmly together with fencing wire — you'll probably need three or four ties along the entire length. The simplest way to do this is to poke two lengths of wire through, twist them together, then insert a small stick and twist it round and round and round till the wires are twisted tight.

The holes where wire pokes through will be used over and over again for more ties.

Now jam in a few 30 cm (or whatever width your walls will be) bits of wood. These are to keep the forms evenly apart.

Check the forms are level and vertical, using your spirit level, either bought or homemade (see diagram page 21). If they aren't, wriggle them around till they are, or give a few taps of the hammer. You may need to trim your spacers a little if they are slightly too wide.

### Step 2

Make a nice moist (but not thin and wet) mix of concrete — six gravel, three sand, one and a half to two cement, plus water. Pour about 10 cm into the formwork. Now arrange your first row of 'good' stones on this, flat sides facing the formwork. Get them as close together as you can, but don't let them quite touch, and wriggle them firmly down into the concrete. If your stones are irregular, you may need to wriggle a few small stones between their curves to minimise the amount of exposed concrete.

Try to lay the rocks as they'd lie if you were building without formwork. In other words, don't balance a rock on a point and make them lie naturally and firmly. Remove the spacing sticks as you go, but leave the wires in place.

Now fill in behind with 'uglies', as many as you can cram in. The more 'uglies' you add the less concrete you'll have to use, and the cheaper your walls will be. Don't have one rock quite touching another though — they should all just float in the concrete.

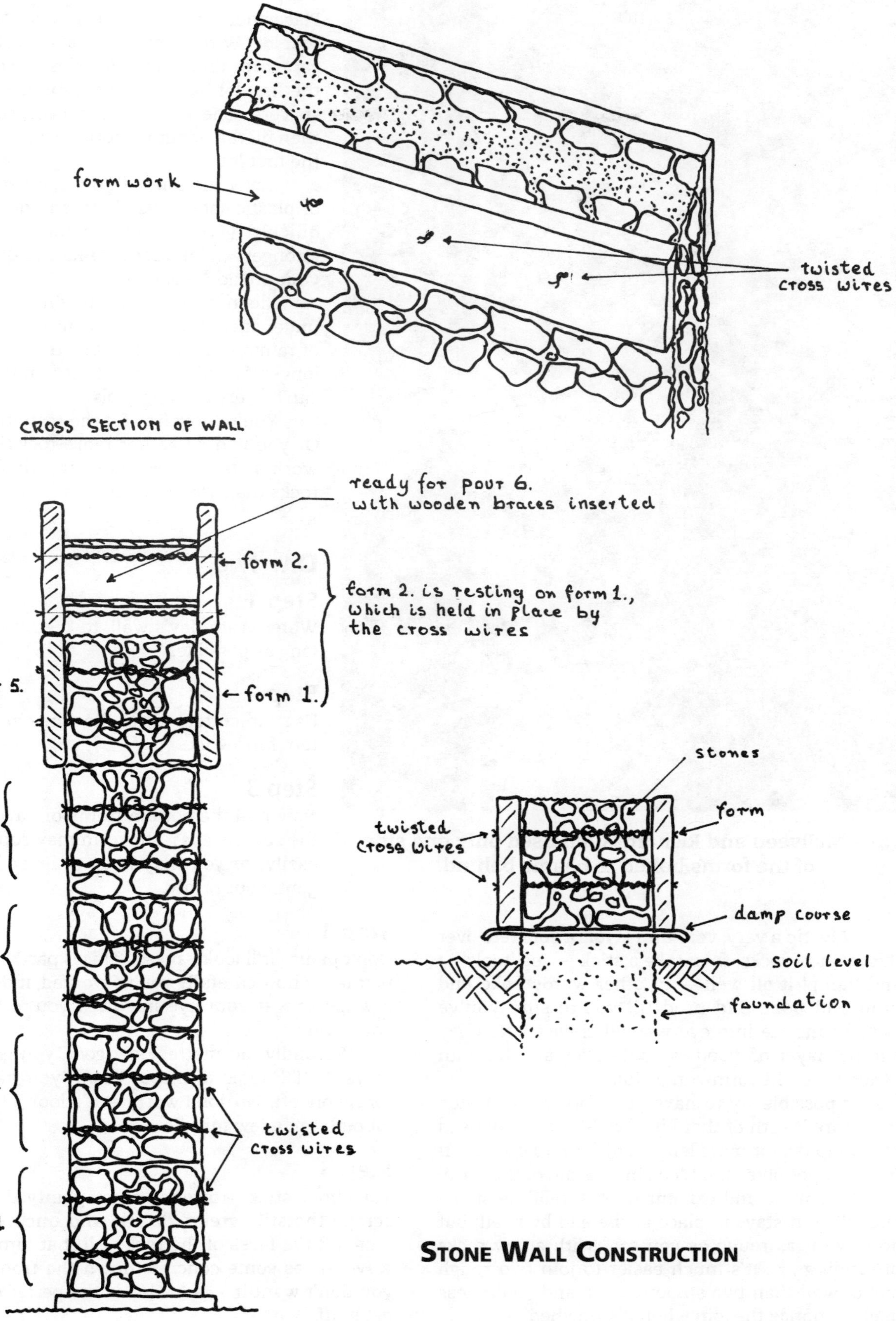


**Stone Wall Construction**

**Sue McIlveen and Richard Allen used only one side of the forms bolted to beams behind.**

### Step 3

Place another lot of formwork on top of the first lot. Wire it onto the first form, then insert wires through the form as you did with the first lot of forms, jam in more wooden spacers. Check it's level and vertical again, then fill that length of form just like you did the first lot.

Cover the top with old blankets, tarps, or plastic sheeting so it doesn't dry out too quickly (slow-cured concrete is much stronger) and in case it rains and dilutes all your precious concrete.

Clean all tools thoroughly; put them under shelter of some sort to keep the dew or rain off them; oil any bits that might last longer if oiled . . . cement dust and concrete can be very hard on tools.

You're now free for the rest of the day. Or you can go gather more rocks. (Stone work always takes about ten times more rocks than you thought.)

## DAY TWO

### Step 1

Water yesterday's wall, to keep it moist as long as possible.

### Step 2

Use wirecutters to cut the wires in the bottom formwork.

### Step 3

Now pull the bottom formwork away from the stone walls. The form may come away easily, or you may have to lever it away, gently but firmly.

Now tip a very, very thin layer of concrete over this layer of rocks, then take a stick — see 'tools' — and 'tamp' it all well — in other words poke and prod and poke and prod till you're sure you've forced concrete into every small crevice. Now lay another layer of 'goodies' and uglies and fill with concrete till the formwork is full.

If possible try to have your formwork stretch the entire length of that bit of wall. You may need to join up two or more lengths of form to do this. If this isn't possible, don't 'box in' the end of the form.

The stone and cement work should be stable enough so it stays in place at the end by itself; but do leave it as rough as you can, with a few rocks protruding . . . it's much easier to join two rough bits of wall than two smooth ones, and you're less likely to notice the join when it's finished.

### Step 4

Don't panic. It'll look a mess. This is partly because no matter how carefully you've worked, it does take a while to 'get your eye in' when you're building with stone.

Secondly, no matter how good you get, it always DOES look a mess when you take the formwork off. No don't worry — it'll look a heck of a lot better in a few minutes.

### Step 5

Take your stick, trowel or other pointy thing and scrape the still 'green' (brittle and uncured) concrete off the faces of the rock. All that tamping always forces some concrete down the front where you don't want it — and this is the perfect time to get it off.

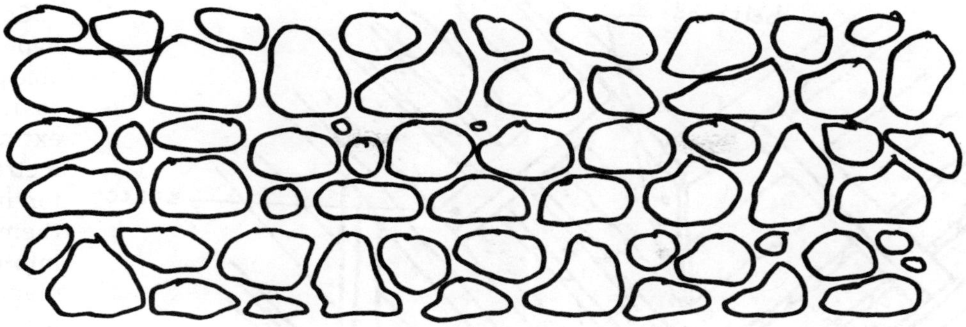


You can also brush the surface firmly with your wire brush and this has the added advantage of gently raking out the joints, to give you a more professional looking finish.

DON'T wait a few days to do this. The concrete will 'cure' and get harder and harder and if you wait too long you'll need a chisel, not a stick or trowel and will possibly remove more of your own flesh in the process.

## Step 6

Water the concrete thoroughly again to keep it moist.

## Step 7 (optional)

This is also a good time to take a small batch of concrete (not mortar — it won't match) and fill in any holes around the rocks — and there are always some. If you want to make the concrete more pliable for this stage of the process add a splodge of detergent (no, I don't know how much as I never measured. Say about a tablespoon per bucketload).

This 'tidying up' stage isn't absolutely necessary — in fact it's one that I never got around to most of the time. I wish I had.

This 'pointing' really does make the stone work look far more professional. You can 'point' at any stage of the building process . . . or even 20 years later but if you wait more than a few weeks the new concrete will be a different colour than the old stuff, and your 'points' will look too obvious. They'll eventually weather down to the same colour after a few months.

Owner-built houses are very rarely ever totally finished — it's all too easy to leave a bit of work for later — and 'later' very rarely arrives. Maybe this is because people who build their own homes tend to be creative and creative people are always going on to the next thing to be created, and are not really keen to do the last bit of boring stuff from the job before. Anyway, if you can bear to 'pointing' really is best done now.

Note: if you are having stone inside instead of a smooth concrete wall, DEFINITELY point as you go along. Firstly, for appearance but secondly because holes around stones are the most wonderful hiding spot for spiders. A house with a stony holey wall is a paradise for spiders.

## Step 8

Cut off the wires as close as you can to the formwork. DON'T leave them sticking out — they'll scratch your legs (or even more tender places) and may rust, which can leave stains on your nice new stonework. Throw them in the bin, not on the ground, or you'll still be picking them out of the grass 20 years later.

## Step 9

Take the formwork you have just taken off the first rung, and place it on top of the second rung of formwork. Use wire to attach it to the formwork below (again, reuse the same holes over and over) and then wire the two formworks together, as you did before. Place the spacers in the forms as you did before, and check it's all vertical and level.

This is the magic of formwork — the bottom form stays attached to the wall by the wires holding the two forms together. Then you just attach the next bit of formwork to the one below.

## Step 10

If the surface of the last layer is very dry, or if more than 48 hours have passed, paint on Bondcrete or some other cement 'glue', so that the two layers bond together perfectly. If the wall below has stayed covered and moist however, you don't need to do this.

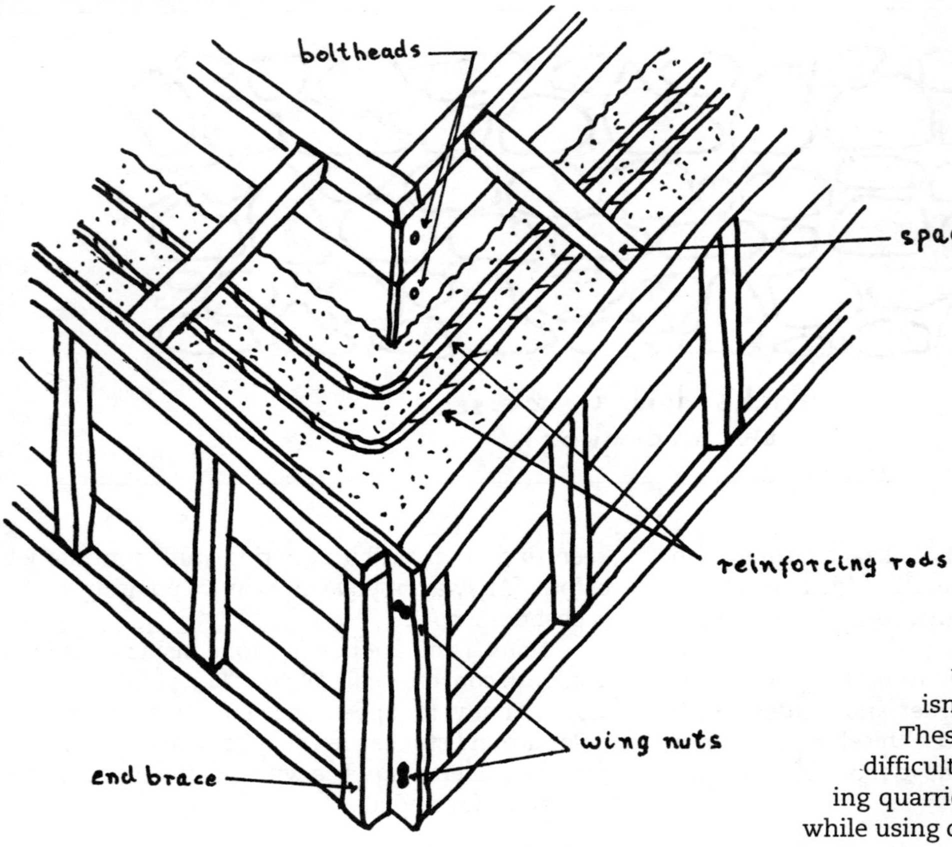


**CORNER FORMWORK**

Corners are perhaps the most vulnerable part of a building, so it's worth taking extra care. Get two metre lengths of reinforcing rod, bend them in the middle into right angles (see diagram at left) and lay a couple every 250 mm or so as the wall rises.

If possible, cornerstones should also have a right angle. In a few places we cheated, and butted two rocks together, meeting at the corner . . . this does not look good and isn't as strong either!

These corner rocks are the most difficult to find unless you are using quarried stone. It may be worthwhile using quarried stone just for these corners.

### Step 11

Fill the third lot of formwork the same way you did the previous two.

### Step 12

Leave the second and third rungs of formwork overnight. Cover as soon as they are done to slow down the concrete curing. Clean tools!

### Step 13

As soon as the concrete is firmish, water lightly to keep it moist so it cures as slowly as possible.

## DAY THREE (AND FOUR AND FIVE AND TWENTYNINE)

Take down the bottom rung of formwork; place more formwork on the bit that's left . . . and keep on going . . .

### Corners

When you get to a corner, wire the forms together at right angles. If you are going to build a lot of corners, it's worth making special corner forms that bolt together.

### Placing doors and windows in the formwork

Fitting doors and windows into formwork is as simple as can be . . . but it's a heck of a lot easier doing it than writing about it, as it involves giving a twist that way or this way, and a bit of temporary formwork over there . . .

### Doors

When you place the first two forms down, stand the door frame up in the middle of it, exactly where you want it to go. Keep it in place either by jamming a stick or six on either side of it, or using the same wiring methods that you used to keep the forms apart.

Make sure it's vertical and straight. Most of our doors aren't, quite. This causes problems later, as doors jam on the floor and need to be planed to fit the odd slope of your doorway.

Now you need plywood and gaffer tape. Cut sections of ply to make temporary formwork on either side of the door. Keep them in place with gaffer tape. I can't really explain how to do this in words, just trust me — once you get fiddling, you'll find it's, well, fiddly, but quite possible. Just be a bit gentler when you're tamping around the temporary formwork.

Don't hang the door from the doorjamb till after the walls are finished, otherwise it'll get stained with concrete. And as soon as you take the formwork off, do wash the concrete off the doorjamb. The longer you leave it, the worse the stain will be. You can also tape plastic or some other covering over the wood that will be exposed so that it doesn't get concrete on it.

We didn't ever bother with that much — our doors and windows were second or third hand, and pretty scratched about anyway.

Basically enough of the doorjamb needs to be embedded in concrete to keep the door stable — and it does need to be pretty stable, as doors are under a surprising amount of stress when they're banged, slammed and generally flap about.

One option is to nail reinforcing wire to one edge of it, so this is embedded in the concrete too. It will greatly help stability, especially with doors that are opened and closed a lot.

## Lintels

All doorways and windows need a lintel, unless they are going to go all the way to the roof line. Lintels can be made of reinforced concrete, or giant hunks of wood, or long thick rocks if you are sure they are thick and strong with no cracks and won't break. See below.

## Windows

The more windows you have, the less stone wall you have to build! On the other hand, you lose a lot of winter warmth through glass and gain a lot of summer heat. But on the other hand (yes, I know this is three hands) natural light is a wonderful thing to have in a house . . . and you can always build a pergola and plant deciduous trees to shade the windows in summer and to keep in the warmth in winter, see page 53.

If you want a floor length window, treat it just like a doorway. If you want your windows at any other level, just place them on the formwork when it's at the height you want for your window.

Do NOT place the glass in the window frames till after you've finished the walls, if possible. Firstly, it might break — which is both expensive and dangerous. Secondly, the concrete will drip over the window and you'll have to clean it — probably with something sharp bladed. This is fiddly, time-consuming and you can cut yourself. The concrete also inevitably leaves scratches on the glass. (You can smooth these over with jeweller's rouge. I have promised myself for years that one day I will do this . . . )

Most of our windows came with glass in them. I think we got the last of the concrete dribbles off the year before last . . .

If at all possible, build your window frames in before you put in the glass!

Now all (!) you need to do is build temporary framework around the window frames, the same way you would for a door.

Note: If you look at old stone buildings, many have attractive stone featurework around the doors or windows. Our building was too rushed to bother with these; I wish we had . . . and am almost (almost) tempted to build another few stone rooms just to try them.

## Loadbearing lintels

A lintel is a load-bearing slab above a door or window. It extends well past both ends of the door or window. Basically it's bearing the load that otherwise the door or window would . . . or wouldn't. It would either bend, crack or your wall might fall down. Also, the door and/or window will refuse to open or close once they are bearing the weight of the wall above them.

So build a lintel.

If you can get hold of lovely massive timbers, like bridge timbers or home-adzed tree trunks, you can use these as lintels. You can also use long rectangular rocks — but you must be SURE that these are strong enough to carry the load, and don't have any cracks or fracture lines.

You can also make rock-arched windows; I have never done this and it isn't covered here.

Our lintels are made of reinforced concrete. DON'T add rocks to this bit of concrete — you may weaken it. If you want a decorative stone effect, use 'facing rocks' that don't go very far in.

What you DO need is reinforcing. We managed to gather the thrown-out bits of steel bars that were used to reinforce a sewerage tunnel. (Our door and window openings are probably the best reinforced in Australia.) We used simply masses of long lengths of strong steel.

If you can't find an old sewerage tunnel to scavenge (unlikely) you will need to buy reinforcing mesh. Work out what size you need and get it cut before you buy it — this saves a lot of messing about with bolt cutters.

This is one area that MUST be checked by an engineer, to make sure the lintel really will bear the load you're going to expect of it. The amount and length of lintel and reinforcing will vary according to the size of the door or window and the height

**The windows extend all the way to the top of the wall so reinforced lintels aren't necessary in Sue and Richard's house.**

and weight of wall above it. So ASK!

A lintel should extend at least half the length of the window again on both sides.

## Doing without lintels

Many amateur builders simply decide to avoid the whole question of lintels — their windows extend all the way up to the roof, or they place windows above their doors that extend all the way to the roof, or place wooden or other light panelling above windows and doors instead of heavy stone walls.

Many home builders decide to have floor to ceiling windows and French doors. This basically means that they are building wide pillars of stonework between stretches of window. This is perhaps the easiest way to build stone walls without fussing too much with the formwork, and the effect can be stunning.

I think we'd have done this if we could have foraged or afforded floor-length windows.

## Finishing the walls

If you have decided to have a smooth interior wall, as soon as you take the interior formwork away you'll find that you haven't — you have lumps and holes instead.

This needs rendering or plastering. I just mixed sand, water and cement, took a trowel and wiped them over as smoothly as I could, which wasn't all that smooth, then painted them.

In retrospect I would have had the kitchen plastered — it would be much easier to clean. (When I scrub it now I tend to take the paint off the tiny protrusions.)

A final word of warning too. If you tamp down too hard in the formwork it will bulge and, as you DO have to tamp fairly hard, it almost certainly will bulge. This means your inside walls may be a series of gentle bulges — not too obvious, but not exactly flat straight either. I rather like these, but that probably just says something about the lack of precision in my character. If you don't like subliminal waves in your walls, you will need to plaster them. (They can make hanging pictures a bit of a challenge).

## How to plaster walls

Dunno. Haven't ever tried it. But I have seen books on it at the library, even if I've never opened them. Just follow the directions, she says blithely, and I'm sure you'll be right . . . or look up 'plasterers' in the *Yellow Pages*.

## Cleaning the stone wall

By the time your walls are finished your rocks will be streaked with dribbles of concrete. This looks messy; it also obscures the beauty of the rocks — one of the lovely things about building with rocks is that every one is different. I spend hours even now just gazing at the rocks in the walls, looking at their different planes and colours. I'd hate to live with walls that were all one colour now . . . nothing to gaze at and contemplate. Some of the worst stains will knock off quite easily with a tap or two of your scutching hammer but there will be thinner dribbles that require chemical dislodging.

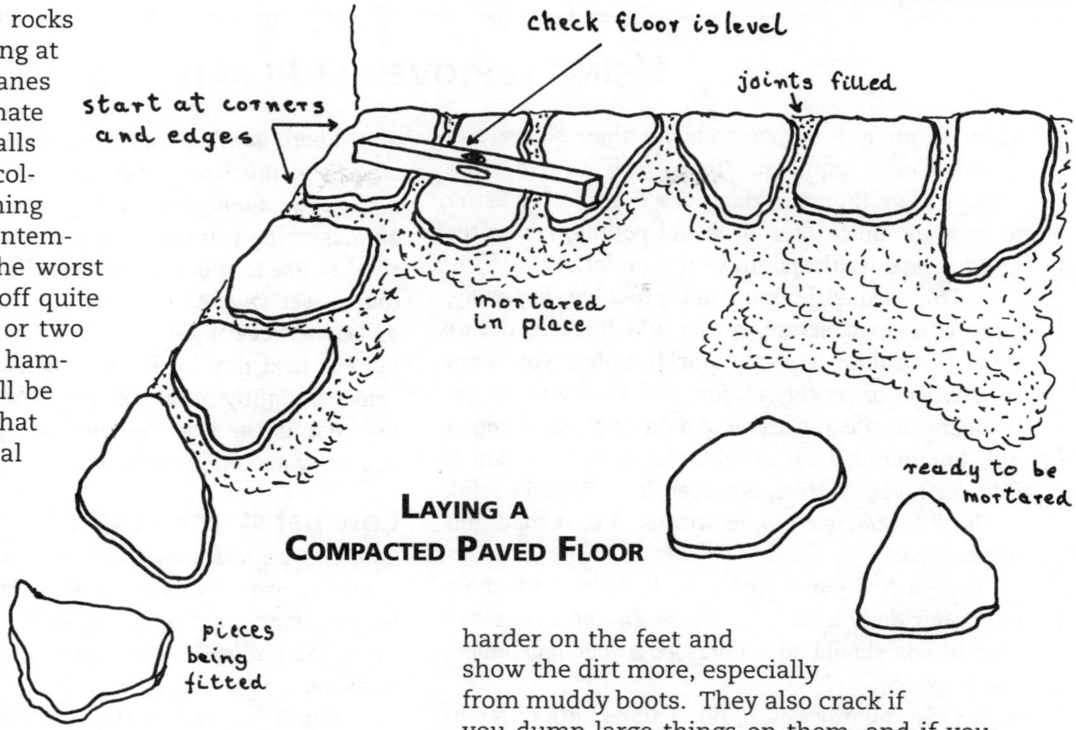


LAYING A COMPACTED PAVED FLOOR

Take gloves, long sleeves, long trousers, plus goggles and a bottle of hydrochloric acid from the hardware store. Check if this has been diluted already with water — concentrations vary. Some can be used straight — some diluted by 20 parts water to 1 of hydrochloric acid. If in doubt, ASK. Always pour the acid into the water, not the water into the acid, to minimise splashing of the acid.

Now wet the rocks thoroughly with water, using either a hose or sponge depending on if they are indoors or outdoors. Scrub the acid onto the rocks using a nylon brush — not bristle or wire as the acid will eat them away. No, it won't eat away all the concrete between the rocks — but it will clean the rocks. Wash them thoroughly afterwards, working top to bottom with a damp sponge to remove the last bits of gunge.

If your nose or eyes begin to sting, STOP AT ONCE, leave the room and splash your face with cold water. Don't go back to the acid till the stinging stops and do seek medical advice if it doesn't or it's severe . . . and be very, very careful of splashes in the eyes.

It should go without saying that pets and kids should be kept well away during this process.

## Floors

Most of our floors are concrete slab, covered either with cork tiles or vinyl tiles, with rugs of various sizes and makes on top. If we build another bit, we'll probably do the same thing — easy to clean, hard wearing and don't show the dirt.

We also have some areas of tiles: these are harder on the feet and show the dirt more, especially from muddy boots. They also crack if you dump large things on them, and if you plan to live in your house for a few decades the chances are that someone sometime will drop something . . .

Several friends have slate tiles. They look fantastic, but also seem to show muddy prints and feel hard on the tootsies. But if they're in areas where you won't be doing a lot of standing — say a study — and where the mud of your boots will have worn off before you get there and you can afford it — go for it!

## Compacted paving floor

This was laid down directly onto a sand base on top of plastic, with fine sand to fill the crevices and a sealer painted on top. It was fast and cheap (which mattered a lot at the time, as both time and money were scarce) but has never really been a success. Despite the plastic and the sealer they do feel dampish, don't ever really warm up and have gathered stains over the years. Wouldn't do it again.

There are a few floorings also worth mentioning:

- a mix of one part sawdust to one part sand to one part cement, mixed with a little colouring (this is cheap and can be bought from the hardware store) laid over the concrete slab. This is much softer than concrete, and if you trowel it with a metal trowel till perfectly smooth it can be polished as you'd polish a wood floor — different effect, but not bad. But on the other hand, not so wonderful that we decided to use it.

## Using leftover concrete

*Most people buy the concrete for their slab ready mixed from a supplier. It's a good idea to order slightly more than you think you'll need. It's better to have too much than a bit not poured, where the join can be a weak spot in your foundations.*

*This means though you'll probably, hopefully, have leftover concrete, and you'll have to decide what to do with it pretty quickly unless you want the supplier to get rid of it for you. You can even get leftover concrete at the end of a day's stone laying: a half barrow or such without formwork to slap it into. Leftover concrete, however, is ALWAYS useful.*

- *Use it to concrete around drains in your road that might erode.*
- *Keep a few paver forms on hand and fill them with your daily leftovers. At the end of a couple of months you should have amassed a good few pavers (see page 66).*
- *Lay out a garden path; fill sections with different colours of concrete, with jointing compound between them (see page 15). You'll end up with a crazy paving path.*

### Cow pat pavers

*These are useful when you have a quarter of a bucket or less of left over concrete.*

*1. Add a small amount of colouring, either the same each time for a neat and elegant look, or different colours for a touch of craziness.*

*2. Fill both hands with a good large splodge of concrete — as much as you can.*

*3. Splodge it down on top of your mound of sand (I am assuming you will have a convenient mound of sand to use to mix the concrete). It will look like a round, wet cowpat.*

*4. Leave it overnight.*

*5. The next day lift it up. The bottom should be smooth, slightly sandy and gently curved. This bottom side is the side that will be uppermost when you use your cow pats as pavers.*

### Cow pat stepping stones

*Cow pat pavers make excellent stepping stones wherever grass is wearing out because of overuse but you don't want a formal path. Dig them in just below ground level, so you don't crunch the mower on them.*

*Cow pat stepping stones are also good if you decide you don't want a grass lawn and plant it out with strawberries, Corsican mint, pennyroyal, woodruff, wild flowers, or predator-attracting flowers. Stepping stones mean you can walk through the area without crushing the plants.*

### Cow pat paths

*If you are making a cement path, cow pat pavers can be dropped into the wet concrete, just as you would rocks.*

- flat stones set in concrete. This looks stunning, especially with colourful mats over it; is very cheap if you have a source of rock; and can be washed easily. It is also tough on feet and you must be sure that it is quite even or you may trip. It's probably best not to use it in the kitchen, firstly because of the tripability factor and, secondly, because it does stain if you spill things on it. Not to mention that there are no second chances with dropped crockery.

### How to lay stone floors

Actually it's pretty easy (although heavy and takes time). You need:

- thinnish stones with one totally flat surface. This isn't as impossible as it sounds as most stone 'splits', leaving one flat side. The rocks don't have to be the same size either in width or height, but they can't be too thick as they have to balance on the reinforcing. Most stones need to be at least three inches (75 mm) wide but if you are using a very strong stone like slate they can be much thinner. Watch out for dips and lips — floors DO have to be even, or you'll stub your toes or worse.

The rocks don't have to be all the same thickness, as they'll be 'bedded' into sand or concrete. The wider the stones the better the effect, but don't break your back. Have lots of small flat stones too for filling in awkward spaces.

- a length of wood almost as wide as your room.
- rubber mallet (this is not absolutely essential, but it helps).
- plastic.
- sand — optional, depending on method.
- mortar (not concrete; see page 26). A little colouring (bought at any good hardware store) in the mortar will prevent the mortar looking too white in between darker rock.

Have ALL your rocks ready before you start. This means laying them out roughly in the position they'll be in the floor, to make sure you have enough. You can also mark off one part of the floor and do that first. Then another and another but the disadvantage of this is that you may not get a good seal between each piece, and the concrete in each portion may weather at different rates and look slightly different shades. You may also not get an even looking pattern. If possible, collect the rocks and go for it . . .

Prepare the site as though laying a concrete slab: dig the foundations, lay down plastic and reinforcing mesh. See also 'Laying rock paving in concrete and mortar ' on page 59.

You can either lay down a slab then lay your floor on top of the slab, or mix the slab (using concrete) and lay the floor as you go, then fill the cracks up with mortar later.

If possible get a friend who will just mix mortar all day (the best way to find a friend like that is to do the same for them when the time comes).

Always lay stone floors around the edges of the room first; this way you can use the edges to keep the centre of the floor level.

Mix one load of mortar; splodge it down in one corner. Don't spread it flat — let the rock push it down, otherwise you may end up with air gaps under the rocks, and the rock may later crack if it's unsupported.

Lay the rocks in it according to your previously determined pattern; then run the long bit of wood over them to make sure their surface is totally level; if not, tap them down with the rubber mallet till they are. If they are too low you'll need to insert more mortar. Tap each rock with the rubber mallet after it's laid. If it sounds hollow there are air gaps underneath and it needs to be bedded down more firmly.

You can either fill in the gaps between the stones with mortar now, or wait till the whole floor is laid and fill them in then.

By now the next lot of mortar will be mixed — so keep going.

If rising damp isn't going to be a problem, that is, your floor is outside or in a shed or it hasn't rained for a decade and building inspectors won't object, you can lay the floor on sand then sweep more sand in between the crevices over several days as the sand compacts.

Do prepare food and drink the day before, so you can grab it as you go.

## If you can't find rocks . . .

Make your own paving stones instead, and pave your floor with them. See 'Making Your Own' on page 66.

Some friends laid down rounds of hardwood instead of paving stones. It looked good, though not as good as paving stones — rounds of wood leave bigger gaps of concrete between them than irregular stones. Then the wood shrank, even though they'd used year old rounds of wood. So they mixed up glue and sawdust and filled in the cracks. They coated the rings of wood many times with linseed oil. I suppose you could also use an Estapol floor paint.

# FOUR

## Bryan And The Second Storey

THE marriage was finished before the house was — like a good many houses and a good many marriages. Building a house together can be one of the most testing activities a couple can undertake.

Why? Well to start with, you both get physically tired and crabby when muscles ache. Money is tight because you are trying to stretch it a few metres further than it will go; and because you are working so hard you can forget to make time for things that are simply fun . . . the sort of happy memories that a good marriage is stuck together with.

But I think there is another major factor too. Many women — me included — need a home in a way that many men don't understand. Our living space is desperately important to us. (It may be significant that I know of many houses built by single women, but only one built by a single man.)

This imbalance in NEED for the house can surface in many ways — a growing resentment at the not quite so great commitment of the other partner, a feeling of dislocation the longer the house stays unfinished. Some men, too, speak of a feeling of being 'trapped' when they put on the front door, or the front gate. All these will put strains on a partnership but they can also make it a heck of a lot stronger too.

I don't think my first marriage broke up because we (almost) built a house together. I think Bryan and I are closer because of the things we have done together . . . house building among them. But if there are problems in a relationship, building a house together is a great way to pinpoint them.

So my first marriage ended; Edward and I lived in what was still a semi-converted machinery shed, three small rooms plus an unfinished bathroom not attached to the rest of the house, with a small, badly-installed solar power system for lights that soon went kaput. I started writing to support us, plus cooked in a local restaurant at night till in a few months the writing began to pay (a story I won't go into here).

And then I met Bryan, which is another story that I won't write here either.

Bryan might have spent his whole life studying how to live here. He'd first qualified as an electrician, then an engineer; had worked on the Apollo moon project with NASA plus other major projects since.

Sometimes I think our relationship really became cemented the first breakfast we ever had together at his place in Canberra. I looked out the window and commented that the sunny day would be good for my batteries, and Bryan looked deep into my eyes and exclaimed, "But darling, why didn't you TELL me you had a solar power system!"

We flourished from then on.

Bryan took charge of the power system; revitalised and extended it in ways I understand for ten minutes then forget; his most beautiful addition is a water wheel, which sploshes around generating electricity in winter or during long rainy stretches.

As soon as Bryan and I had bought the property after the divorce we moved back down there together. In the meantime we arranged for the framework of a new high-sloping roof to be put on. This replaced the flat 'shed' roof that was already beginning to leak. The new roof also gave us a large 'attic' space, with a narrow staircase jammed up along one wall of the laundry.

We moved in before the attic was finished of course. Bryan was still nailing on lining boards months later and I don't think we have ever quite got around to screwing down all the tin properly. Ah, excuse me, Bryan says, we have — Bryan finished it last year, abseiling down to do the last of it — it's a VERY steep roof.

So we had a house. Not a big house and the rooms were inconveniently small and there was nowhere we could have six people for dinner unless someone had their nose in the sink and I sucked in my stomach as I collected the plates so my bum didn't get burnt on the stove.

We lived like this for years, and were happy, if cramped.

As money became available we built a new room for Edward out the back (in wood, not stone) and then Steve Kidd, the Braidwood master wood craftsman, demolished the old wooden internal walls (I think the stone internal walls are immovable) downstairs so we had one big (well, bigger) room plus a dark, cold larder in the middle of the house. The larder is always about ten degrees cooler than the rest of the house. It only warms up when the temperature outside is over 40 degrees for at least ten days, and then we have to remember to leave the door open at night so it cools down.

The house was still too small, and Edward's friends were growing larger. One weekend I realised that there wasn't even floor space for half a dozen of them to lie down at once — and there wasn't enough shelf space for all our books either, much less the ones that were sure to accumulate over the next few years.

So we extended again.

It isn't easy to extend when you have a steep hill behind you, a sharp drop in front of you, and thick stone walls all around. We finally decided that what we needed wasn't more rooms — well, maybe one more room — but a larger central living area.

The only way to do this was to build a sort of wrap-around room around the front part of the house, taking out the windows and the stone windows sills below them, to make two giant arches leading into the new bit. (See ground floor plan on page 76.)

This was not going to be easy.

So for the first time we consulted an architect, hoping he'd have a more creative solution. But he just agreed with our original idea — without really answering the problems about how the bits would be joined on — and drew up plans for us, which took five months, plus many phone calls, and was expensive. (I'm pretty sure that will be the last time we use an architect.*) And in fact Ray Spriggs, the local builder, was the one who worked out how to get around the tricky bits, like joining a wooden addition onto stone walls, and we decided we didn't like the proportions the architect had proposed, or the materials he suggested, so the whole thing was a waste of time, except I suppose for convincing us that if he thought it was possible, then it probably was.

So we went ahead.

Ray did most of the work, with Bryan as builder's labourer. My role was just cups of coffee and chocolate chip biscuits and dashing up to Braidwood for more nails and saying, "No, no, I think it should go out another foot there," and changing my mind half a dozen times about whether we wanted floor length windows in the front.

**Bryan's most beautiful addition is a water wheel, which sploshes around delightfully in winter or during long rainy stretches.**

Steve Kidd built the windows — gorgeous great wooden things that slide across like penguins on an iceflow. Anyone who knows anything about carpentry makes a beeline for them — they really are a work of art, as different from aluminium windows as chewing gum is from a double chocolate peppermint cake with almonds.

Steve also scavenged some old timber from the Sydney wool stores that accidentally burnt down after they had been classified as heritage buildings. (Strange how often that happens.) They looked a mess — stained grey old wood. But he smoothed them and reformed them and turned them into what look like giant beams holding up the arches, beautifully grained and puckered with age.

*No, it's not — we're doing it again a year after I wrote this!

**The new roof replaced the flat 'shed' roof that was already beginning to leak and gave us a large 'attic' space.**

In reality the arches hold themselves up very nicely, thank you, with all that reinforcing put in so many years before. But the beams do look stunning — I must run my hand over their curves and striations at least once a day.

It was NOT easy drilling out those stone walls though. "No problems," Ray had said blithely, but he'd never come across concrete quite as strong before, not to mention the granite rocks. Those walls had been built to last. Everything in the house — even upstairs — was covered with a fine film of greasy concrete dust, and I think all our softer tissues were still vibrating three weeks later. But it was worth it.

So we didn't really build the last bit, but we guided every stage of it. It was our vision of a room that would fit the way we lived our lives, and a bit of our sweat went into it (a lot more of Bryan's than mine). It feels like ours and it is beautiful.

In winter the sun streams through the long tall windows and warms up the cork-tiled floor and this heat flows into the rest of the house and warms it too.

Even in summer, when the sun is high and no direct sunlight sneaks inside, you still feel you are part of the garden. The red-browed finches peck at the seed ball hanging in the tree about a metre from us as we sit reading, and the lyrebirds scratch at their reflections in the windows in masculine fury at the mirrored interlopers.

It's a good room. It's a room to sit and watch the sunlight and the shadows, to reform the world over breakfast and plan new gardens over lunch. I'm not sure if it's the last room I'll ever be part of building. I hope not. I hope there'll be lots more rooms to come . . . a small guest house and studio, perhaps, the sort you need as kids drift away and then drift back again, bringing their family for visits.

But we shall see.*

*One year later we're about to build again! Another bedroom, plus passageway . . .

FIVE

# Auntie Jackie's Advice On Stone House Design

MOST houses aren't really designed, by which I mean actually thought deeply about. They're just repetitions of what everyone else has, with a few variations — "Oh, let's have the third bedroom at the other end of the house so it can be a study," sort of thing.

Having lived in an owner-designed house for the past quarter century or so and with most of our friends living in houses they designed themselves too there are a few design points I've learned. The most important question of all is — how is your house going to fit into the way you live?

If you are going to live a 9-5 life, mostly getting home after dark and spending the weekends catching up, forget about a giant garden you'll never see. Have a small lighted courtyard with good smelling plants you'll actually USE.

If you plan to work from home, don't just use the third bedroom. Design a really good workspace, either at the centre of the life of the home or away from it, according to preference. I like to work near the kitchen, so I can put dinner on and do the washing up in RSI breaks. But if I had a family that whinged or played rock music I'd move further along.

If you're going to have kids — where will their friends sleep or party, or gossip without you breathing down their necks?

Do you really need a formal living room, or just a great big family space? Will life be lived in the kitchen or the living room? If it's going to be the kitchen make sure it has the best view and morning sunlight.

## Kitchen

**Walls.** Make sure the walls are SMOOTH. Rough walls look rustic but accumulate greasy dust. Kitchen walls should be washable.

**Floor.** Use a comfortable material to walk on such as cork. Slate looks great but leaves you with sore feet. Kitchen floors should also be easily washable, and not look too grotty if you spill flour on them so use a slightly patterned rather than plain surface, again, like cork. Make sure there are no rough edges where you can trip or stub your toe. Some friends have stone paving in their kitchen — gorgeous but you need to know where the odd 'lip' is, particularly if you're carrying a saucepan of hot soup.

**Bench space.** No enthusiastic cook EVER has enough bench space. We have a centre bench in our kitchen that can be moved where needed. Its top also opens out to double the possible bench space. When I'm not cooking for dozens it's out of sight. Make sure the kitchen 'flows': that you can stand in the middle of the bench and reach to the stove or sink comfortably. Also make sure that people can sit in your kitchen and talk to you without getting in the way.

One thing I would adore is another sink. I do a lot of cooking and the rest of the family does a lot of eating — and drinking. It's a nuisance trying to drain the spuds when someone is getting another glass of cordial or washing carrots. If I were designing our house again I'd have a scullery with a second sink and bench space just for washing vegies and dumping dirty saucepans, and where everyone else can get a drink without being trodden on.

## Taps

If you have your own water supply, consider having three taps at the kitchen sink, bathroom and laundry — one for cold dam or creek water, one for cold rainwater for drinking and brushing teeth, and another hot water tap.

## Larder

If you plan to grow a lot of your own food, where the heck are you going to put it when you've picked it? We have a small room in the centre of our house, with thick walls, no windows and a relatively stable temperature. There are shelves floor to ceiling for jams, jellies, chutneys, preserved tomatoes, staples like giant cans of olive oil, honey, pumpkins, sacks of spuds, the boxes of apples, oranges, kiwi fruit, tomatoes etc, all saved from the harvest and that need to last us for a few months more.

Our larder is really much too small, although enormous by suburban standards. But when you are going to eat one pumpkin a week for six months, a case of apples a fortnight, ditto oranges, have four cases of green tomatoes to ripen over winter . . . well, you need a heck of a lot of space. And pumpkins up the bedroom stairs are a nuisance.

**No enthusiastic cook EVER has enough bench space. These benchtops at the Sue McIlveen and Richard Allen house are almost ready to use.**

## Bedroom

If you're going to have kids, make sure your bedroom is away from their living area. Teenagers sleep from dawn till afternoon, and the early hours of your sleep time are their noisiest hours.

Try to have morning sunlight to get you up. Not afternoon sunlight so it's too hot to go to sleep: bedrooms should face east or north — definitely not west.

## Guests

If you live in the country you will have guests — not just for lunch, but overnight too because it's just too far to get to you and back again easily. Guests change as you get older. Twenty years ago our guests were happy to sleep in a tent; 15 years ago they loved the yurt at the end of the garden. Nowadays as we wander to geriatric-hood most of our visitors prefer to have a loo near at hand at night, firm comfortable beds and suddenly we have the next generation of toddlers again too, just when we thought the place didn't have to be kid-proof.

Guests are usually on holiday. This presents problems if you work from home, unless you can take a holiday too every time you have guests. This may mean that most of your year is holiday.

The answer is either to have a study/workshop separate from the house, or a guest-house separate from the house. Which is what we plan to build when we get around to it. Guest houses are also great for independent teenagers, a bit of bed and breakfast or giving workshops on the sex life of lizards.

## Bathrooms

Every standard bathroom I have ever known has a shower where you knock your elbows washing your feet. Showers should be large and luxurious. After all, you are going to spend a lot of (hopefully) sensually-delightful time in them. Make sure the hot tap is one nearest you too, so you don't get scalded turning it on!

- make room for spare towels and toilet paper.
- think of how many towel racks you'll need, and double it. Guests will have towels too, as will (as yet possibly only theoretical) kids. In cold climates it's good to have two towels on the go so each one can dry before you use it again. I like to dry myself with two towels when I wash my hair: my hair dries in about half the time and I don't have to bother with a hair dryer when it's cold.

## Windows

Hot air rises and hot air is what you usually want to expel. Have lots of smaller windows above your regular windows and doors so the air can either escape faster, or escape without causing a draught around your ankles.

## Wall holes!

It is incredibly easy to forget to leave holes in your stone walls for plumbing, electricity cables etc. It's a good idea to leave more of these holes than you need. A simple way to leave a neatish hole is to take a length of black poly pipe the same width as your formwork and wedge these in at appropriate intervals. It's better to have too many than spend hours drilling through stone and concrete!

Surplus holes can be plugged with something easily removable — even plasticine or Blutak — till they are needed. You will need to plug them, otherwise rats, mice, flies *et al* will be sure to use them as a freeway.

We find that it is easiest to have electric wiring for lighting in the ceiling, instead of mucking about trying to run it in cables along or through stone walls. Most of our lights operate via pull switches on long cords hanging down from the ceiling, instead of switches on the walls. This means you use less wiring and minimise loss of voltage. It also means you don't need wiring in the walls.

## Overflows

Beware the overflowing bath! At some time EVERY house has a disaster with an overflowing bath or laundry tap — especially if there are kids. The last thing you want is a good solid waterproof room that slowly fills up with water. Make sure you place a few pieces of poly pipe (see above) at the base of bathroom and laundry walls, or anywhere else there may be a water accident. Make sure that your floor drain (in wet areas) is in the lowest point of the floor and that items such as bath mats and towels do not get put over it. That way water can flow outside with minimum damage.

### CURVED WALLS

*I've never built a curved stone wall (yet). But they do look stunning — before I die I want a stone tower: the sort Rapunzel let down her hair from. I could sit up on top with my rocking chair and yell rude things at my descendants. Curves are also easily built using formwork: you just need round forms. Round forms are best made from metal, not wood (metal bends, wood doesn't). These rounded forms can then also be used for making concrete tanks, small observatories as well as round houses, or curved walls on otherwise straight houses, or small stone towers for the romantic.*

## Fireplaces

Fireplaces use wood. Carrying wood inside drops bark, sawdust, small beetles and the odd spider through the house. When planning a fireplace work out a convenient CLOSE way of bringing in wood. A hole in the wall near the fireplace — like a doggie door — works well.

I love wide fireplaces but most of the heat does go up the chimney, and they can use an enormous amount of wood. We have a slow combustion instead which uses little wood — four chunks can last 24 hours, emits little smoke and has a transparent front so you can see the flames. (Much wood burnt in Australia depletes forest and paddock of good elderly trees that small and large beasties need to live, eat and shelter in, and the smoke is an extraordinary pollutant too. We mostly burn the wattles that fall down over our road and garden (the tops are mulched) and down here in the valley most of the pollutants settle on our place. Our slow combustion stove is also used for cooking and heating

## LET THE OUTSIDE IN

*Houses evolved to protect us from the world around us, to keep us warm or cool and safe from beasties chomping our limbs off while we're asleep. Modern houses, though, usually do this too efficiently. They insulate us entirely from the world around; the air inside them is artificially cooled or heated so you can wear much the same clothes year round; they're deodorised so that many kids are unfamiliar with the scents of roses, ripening tomatoes, or early morning sunlight; the view from their windows is just of other houses, roads and gardens planted with the same sort of grass and shrubs that every nursery sells from Hobart to Cairns. The only tether to the outside world comes from the distorted images of life via TV.*

*If you don't want to spend the greater hours of your life divorced from the world around you, build a house with windows:*

- *long windows, almost to floor level (some of them at least) so that you can see the world from grass to sky without having it chopped off at the knees.*
- *tall windows, ditto, so you can see the moon or birds high above you.*
- *windows that open, so you can smell the outdoors. But do install insect screens too.*

*If you don't want to see too much of the world nearby, build a courtyard or plant shrubbery (See 'Privacy') and create a small private world that you DO want to open your windows onto.*

*People are influenced by the spaces around them. If you close yourself off physically too much from the world outside, you'll close yourself off mentally and emotionally too. And if you really can't bear to open your windows to the world outside, perhaps it's time you moved somewhere where you can.*

water in winter, when the solar hot water may not be enough, if we have a long break of cold rainy weather or lots of guests.

## Stairs

Two storey compact houses are cheaper to build; take less time to build and are best for wiring a home power system because the shorter the distance, the less voltage drop. On the down side, you have to lug stuff up and down them — especially double beds, awkward desks and vacuum cleaners; and there comes a time in most people's lives when due to aging knees, hips, hearts or broken legs, steps are a problem. One solution is to put guest quarters upstairs. But remember that guests often turn out to be your elderly relatives with the aforesaid dicky knees, hearts etc!

Attic spaces are also VERY hot in summer, even with good insulation, as all the downstairs heat flows up into them. If you have an attic (roof space) room do make sure you have plenty of places for the hot air to exit — doors, windows plus perhaps passive extractor fans.

## Modifying a stone house

It is NOT easy to modify a stone house. We've done it and had to live with the noise, vibration and dust in every conceivable and some previously unimaginable crevices.

On the other hand, once you have put the love, energy and sweat into building a stone house, it is every difficult to say good-bye to it. When you are building with stone, either make the house as big as you will conceivably ever want (difficult) or make sure there are convenient exterior doors or floor to ceiling French windows if you want to add a room.

Don't get too carried away by the joy of interior stone walls. They look great . . . but are incredibly hard (no pun intended) to join shelves onto, or even hang paintings . . . though it's not impossible (you can hire a Ramset gun that fires into the stone and concrete or hire someone to do it for you).

A few interior timber stud walls though will make all this much simpler and also mean you can enlarge rooms more easily if you ever want to remove them.

## Marsupial and rodent proofing

Our back door has a thick sheet of tinplate bolted to it. This is to deter wombats who would prefer to sleep on the sofa than in a wombat hole. Wombats can be incredibly determined . . . in fact if you plan to rear orphaned wombats, a stone house is essential — they can dig through almost anything else. A concrete slab is also essential — many a wombat lover has had a small furry head appear through the wooden floor at their feet in the middle of a dinner party from which the wombat felt excluded.

**Wombats can be incredibly determined . . . in fact if you plan to rear orphaned wombats, a stone house is essential.**

Even if you don't intend to be on close terms with a wombat, your house must be rodent proof . . . and fly, mosquito, spider and possum proof.

Make sure there are no crevices anywhere; mortar or putty them up, or if necessary buy a can of expanding foam at the hardware store — expensive and a last resort and I hate to think what it would do in a fire, but it does fill up awkward holes.

Install rubber weather strips on the bottom of ALL doors; all windows should be fly and mozzie screened. Look at your roof and think like a rat or a possum, and fill up every tiny hole you can. If the possum is already in there, disco music and strobe lighting will get them out. Then seal them out.

**Our house has always been a continuation of the garden . . . or maybe the garden is a continuation of the house.**

## A HOUSE THAT GROWS FROM THE LANDSCAPE

*A modern house is mostly built of timber carted from thousands of kilometres away, bricks that are the same no matter where you are, finished off with rainforest timbers from another country. A house built from local material feels tethered to the landscape. Around here there are owner-built houses made of pisé (rammed earth), stone, wood, and mudbricks. All of them are different, built to accommodate the way the owners choose to socialise and earn their living. Jenny's tiny living room is always filled with friends and laughter and fabric; Angela's semi-baronial hall with a fireplace that would roast an ox, but hasn't as far as I know, and a giant table alternately filled with books, newspapers and people; Gille and Victoria's almost medieval thick-walled French farmhouse, Sue and Richard's English-style stone cottage.*

*None of them looks at all alike; all of them sit happily into their surroundings, because they are built with the generosity of the land around them.*

## Things I'd love to have

I would love a small glasshouse on the north side of the house to trap more sun in winter, and so I could grow pepper and vanilla orchids . . . in fact how about a really giant glasshouse so I could grow my own chocolate?

A giant dam for watering the garden in dry times and water storage in case of a bushfire.

A roof that opens over my bed so I can watch the stars and reliably shuts again.

Underfloor heating and cooling — water pipes in the slab that take either hot or cool water through it to warm or cool the house. These were planned in the original house. I know of a couple of houses that have them; they work well, and I really recommend them.

A butler. (Bryan says a maid too.)

## Designing a house and garden to fit together

Nowadays 'a house' is mostly just that — a box to shelter in and keep out as much of the outside world as possible — even peering over the back fence is mostly forgotten nowadays as people focus inwards on the TV set rather than outwards to the neighbours and the world.

Our house has always been a continuation of the garden . . . or maybe the garden is a continuation of the house. Yes, the latter, definitely.

Right from the earliest days here, when I lived in the shed across the creek, the outdoors was part of my living space. That was where I cooked, after the kero stove blew up one baking fire-crazed day (luckily I wasn't bending over it at the time).

I washed outdoors, either soaping myself and rinsing off in a small dish then plunging into the creek (given the creek is only a few degrees up from ice even in mid summer, this required a hardiness I don't think I have nowadays).

No, actually that's not true — part of me still longs to do it. But in those days I had a spare hour or so to bathe every day, whereas now a ten minute shower is so much more time effective. It did feel incredible bathing in the creek, like your body had revved up a thousand times and you could zing up to the clouds and float back laughing. The trick was to jump in once and dash out; the second time your body's prepared for the cold, though you do have to keep moving . . . and afterwards you can take on the world.

So the garden was my kitchen, my bathroom (the loo was a roofless pit dunny among the orange trees), my larder — in those days if I didn't grow it or forage it I mostly did without.

Nowadays we do have an indoor bathroom (I do love indoor plumbing, at least at night — when the rain and wind are howling it is still an almost wonderful luxury to sit down on a sheltered loo). And cooking is done indoors, though I keep planning an outdoor kitchen too, which we will build one of these days — a place to cook on sweltering days when even boiling a kettle adds more heat than you can bear to the kitchen.

But the garden is still part of the house.

It's certainly the place where most of the food

### THE BASIC GARDEN

*My absolute minimum garden would have to have:*

- *a lemon tree, avocado tree (both will survive in warm sheltered spots even in cold climates — our winter temperatures can go down to minus 9° C), one early, one medium and two late maturing apple trees (omit the medium one in fruit fly and codling moth prone areas).*
- *rhubarb and asparagus beds, or scattered through flower beds.*
- *a herb bed with LOTS of parsley and basil planted each year, plus thyme, tarragon, 20 garlic chive clumps, peppermint, culinary mint and eau de cologne mint and a rosemary bush, as well as aloe vera.*
- *a small pond and fountain for reviving me and refreshing the birds.*
- *at least one tall tree (for me and the birds).*
- *something blooming and something scented every day of the year.*

*Actually I am a compulsive grower. If a cruel fairy dumped me into the middle of New York I'd plant apple pips in the paving. I don't think ANY garden would ever completely satisfy me. One of my fantasies is having a whole planet to plant out.*

## USING THE GARDEN TO HEAT AND COOL THE HOUSE

*Our house is surrounded by paving stones with a pergola above them. In summer the pergola is covered with grapes, hops, passionfruit, roses and kiwi fruit. In winter the leaves drop off everything except the passionfruit, but that doesn't matter because the bower birds eat the passionfruit leaves all winter anyway — the passionfruit doesn't seem to mind; in fact I think it grows all the better for its winter prune.*

*This means that in summer the house is constantly shaded; there's an extra cooling effect too from the moisture evaporating from the leaves, and also when we water the plants and the pergola (I'm not talking an imperceptible drop here — I reckon we get a good 15 to 17°C drop in temperature).*

*In winter the sun glares down on the paving and the stone walls of the house, or through the windows to the stone interior walls and the concrete floor. These act as heat sinks; when the sun slides down behind the ridge the walls and floor are still warm to touch, and this heat keeps the chill off the house till next morning — theoretically anyway. After a few cloudy or rainy days it's COLD unless we put the fire on.*

*In summer the same thick walls cool down overnight, and help keep the house cool during the day. At night every possible window and door is opened so the place cools down as quickly as possible. We know we've had a hot spell when the air inside the larder — a thick walled, windowless room in the centre of the house — is hotter than the air in the corridor early in the morning.*

*The house cools down quite quickly, mostly because hot air rises — which means that as soon as we open the doors and windows the hot air flies to the top of the house, causing a 'thermosyphon' that draws the cool air in. Even on totally still days we still have a small internal breeze wafting in from the windows and out through the doors at the top carrying with it good summer scents of rose and curry bush too. Or bushfire and hen manure as the case may be.*

*Other ways of using a garden to warm and cool the house include climbers like deciduous ivy or roses on the wall — excellent natural insulation.*

is stored. We have about 600 fruit trees, most of which need almost no tending — just mulch or other tucker once a year; there are about 160 different sorts of fruit (at last count — I suspect there are more now) including 60 different sorts of apples that fruit from December through to late July (each year I swear I'm not going to plant more apple trees, and I always succumb), plus a lot of 'bush tucker' trees. I'm still learning the climatic tolerances for those . . . even some tropical species survive in this southern climate.

There's never a time when we can't wander outdoors and pick at least half a dozen different sorts of fruit and often a heck of a lot more.

As I write it's spring, the leanest time of the year fruit and vegie wise. It used to be known as the 'hungry gap', when winter stocks were finished and the summer crops were still just sprouting.

At the moment we're picking avocadoes (our place goes down to minus 5° C sometimes in winter but avocadoes and many other subtropicals do tolerate heavy frost if you grow them properly — see *The Best of Jackie French*, HarperCollins), limes, lemons, citrons, grapefruit, mandarins, navel oranges, valencia oranges (the bower birds have just eaten the last of the Seville oranges and the kumquats), macadamias (all six of them — the tree has just started to bear).

The tiny alpine strawberries in the hanging basket in front of my study are almost ready to pick (it will be months before the main crop strawberries are ripe). The loquats are swelling but it's still a month or so to go before we can pick them, and there's masses of rhubarb. Not to mention the first of the asparagus poking its heads up all around the garden — while we do have asparagus 'beds' a lot is simply planted between roses and other flowers and the ferny tops do look pretty with their red berries over summer.

We still have late apples sitting in the larder too — early fruits generally don't last long — the later they are, the longer they last. Hopefully the late July apples (Lady Williams, Sturmer Pippin and . . . blast it, what's it's name? Who cares, it tastes good . . . ). Late apples really seem to concentrate their flavour as they age, just like good bottles of wine (and people too).

As for vegies — leek, lettuce, carrots (about seven varieties, including a red-skinned and a purple one and a pure white one too — very sweet), beetroot, silver beet, spinach, spring onions, broccoli going to seed (we're sick of it) and bok choy *ditto*, giant winter radishes, turnips (mostly use the leaves as a green vegies), parsley, celery, burdock, amaranth. And I'm sure there's more if I went down and refreshed my memory, but at the moment we're gorging on asparagus, asparagus, asparagus and I don't really care what other green vegies are surviving.

I'm constantly shocked by the lack of food at friends' places in town. How can you possibly cook if you have to make a list and buy what you need? Most of our meals here depend on what's tempting in the garden . . . and what's leftover goes to the chooks, who give us eggs and manure the garden for us (though their nightly droppings need to be distributed every few months), not to mention endless entertainment. Who needs porno videos when you have a rooster with afternoon lust in his eye?

The garden is part of the house though in many other ways. It certainly helps cool the house in summer (see below) and warms it in winter. And the house, in its turn, protects the garden . . . many

## BEAUTY VERSUS MORTGAGE

***Life is so extraordinarily short it's a pity not to cram it full of as much beauty as possible. Yet most of the time people blindly accept a bland, boring, third rate place to live: the place where they will spend the heart of their lives, not to mention the single greatest expense in their lives too.***

***A beautiful, fascinating, comfortable house — one that suits your heart and soul — doesn't have to be a sprawling, complicated, architect-designed mansion. It can be tiny — as long as none of it is wasted. How on earth can people mortgage their lives for a brick box with ensuite and green carpet out the front, henceforth referred to as a lawn which has to be mowed every second Saturday?***

***You don't need vast quantities of money to have a decent home. You just need enough brains to actually make a few decisions instead of blindly following what most other people do; the determination to follow up those decisions and about three to five years of hard work, as opposed to the rest of your life paying off a mortgage for something you neither love nor hate.***

of the plants that survive here like ginger and turmeric wouldn't grow without the warmth reflected and radiated from the house. And there's no way we'd have rhubarb all through winter and strawberries in early spring without the warmth of the house either.

The 'outside' also gives us water — the stuff collected in our rainwater tanks; it gives us power — either solar power from the photovoltaic panels on the roof (plus solar hot water too) or the water wheel down in the garden.

Bryan built the water wheel on and off over three years; if it turns all day it gives us just enough electricity to power the house, as long as we're reasonably economical. The theory is that if it's sunny power comes from the solar panels; if it's rainy the power comes from the water wheel. One day we're going to have a long foggy drought with neither sun nor rain and we'll really be stuffed, but till then the system works well.

The garden gives us wood too; we have a wood burning stove which is mostly fed from fallen elderly wattles, though we do buy some longer burning hardwood too. The garden gives us herbs to flavour food, streak my hair, occasionally dye clothes, soothe a sore throat or sunburn, make perfume or cleansers or scent the house . . .

The main thing the garden gives us though is hard to put into words — a sense of joy? A sense that we too are part of the universe?

Humans need growing things around them. That's what we evolved with; that's what we're happiest with. You only have to look at the studies of the effect of green plants on absenteeism and illness in offices or in reducing vandalism in high crime areas to realise that there is something very deep in the human psyche that responds to green and growing things. It's now thought that part of the SAD syndrome — seasonally affected disorder, or winter depression — isn't just lack of sunlight. It's the subliminal lack of the scents of plants too — grass and flowers and green leaves.

Humans are seasonal beasts too. I feel that some deep part of me is being starved when I'm away from home for more than a couple of days in a place where I'm divorced from the seasonal changes — in a hotel or lecture theatre. Every day here is part of the progression of the year. At the moment the day by day opening of buds and flowers, small shoots in brown soil, mulch starting to decay after its winter-long rest on the soil, the sudden re-emergence of a host of lizards on the first warm day, the swallows returning, the disappearance of one of the pair of wedgetails as they begin to nest, the first of the asparagus and the last of the mandarins.

Humans need this. They need to pick their own food sometimes; to have the sense of security of fruit ripening around you, of vegies in the ground, of rocks and trees that could be used in a pinch for shelter or play. And it's this which is the most important thing that our garden gives us.

## Subtropical and tropical houses and houses in humid areas

We live in an area that's bloody cold in winter and bloody hot in summer, except at night (it's wonderful in autumn and spring . . . and there comes a time when even drought and sleet become part of the pattern of your life).

Our main aim in winter is to get as much sunlight as possible into the house, then store it so the house stays warm at night . . . hence the solid walls, floor, large windows. In summer we try to shade sunlight from the house all day with trees, vines and blinds, then open it up at night. We also have wide eaves — in winter the sun is low in the sky and reaches in our windows; in summer the eaves block the higher-angled sunlight.

If we lived in a hotter climate though, our house would be very different — a lighter wooden structure that cools down QUICKLY; lots of cross ventilation (down here they're called draughts); decks with slats rather than a solid floor, lots of small windows above doors and main windows.

## Soundproofing the house

I once read somewhere that if men are frightened they tend to hide under something; whereas if women are scared they tend to climb on top of something: which just goes to show that our behaviour from 100,000 years ago is still with us.

Humans were once scared of many things — we are still, of course, but different things. Not the tigers in the dark behind us but the mugger lurking in the shadows . . . and we are still happiest when we know, consciously and unconsciously that we are secure. I always get a tiger over the shoulder feeling in most hotel rooms; the sense of someone, I don't know who, not very far away, above, below, behind . . . and who knows if they're friend or foe?

Homes should not just be secure — they should feel secure. If your house is buffeted by traffic noises, you'll soon get used to them — consciously — but your unconscious won't. People who live with constant noise are simply more stressed.

There's a corollary to this too. The more unpleasant the noise you live with, the more you block out sounds — not just brake grinds and accelerator roars but the sounds of trees and the crickings as the roof expands on a hot day and rain on the roof or pavement.

Noisy blocks of land are usually cheaper. Don't be tempted. Apart from your quality of life you'll find that no matter how much energy and money you put into your house the resale value will always be lower because of the extra noise.

The choice of where to live affects your quality of life more than almost anything else. As far as possible, don't compromise. On the other hand, SOME human noise is inevitable unless you decide to live out in one of the more inaccessible parts of the Nullarbor. The following are ways to help reduce it.

### 1. Insulation

Insulation cuts down noise as well as retaining or keeping out heat. Use as much as you can! I once heard of a radio station on-the-cheap that soundproofed its studio by gluing cardboard egg cartons onto the walls — the bumpy, egg holding side is what you need. Every amateur, home recording studio in the seventies — and for all I know they still do — used this trick. Doors and windows were rendered useless but in the days before pink batts it was quick, cheap and wonderfully effective. It worked a treat, though it is possibly a little unaesthetic for home use.

Basically anything you use to keep the house warm will help cut out at least some noise — floor length, double material curtains, pelmets, carpet, lots of shelves.

Carpet, shelves — particularly bookshelves — and thick upholstered furniture will also tend to 'soak up' noise. A 'modern look' bare restaurant will be much more noisy than one with heavy chairs and tables, lots of pot plants and the odd chest of drawers, plus paintings or wall hangings to help stop sound being reflected off the walls.

But ANY insulation will help keep down noise — including floor to ceiling bookshelves — full of good thick books, naturally, the best way to insulate a house or study I know!

### 2. Double glazing

Double glazing also cuts down noise enormously . . . just try shutting the window on a noise and seeing how much softer it is; well, double glazing will cut out about that much noise again or even slightly more.

If you can't afford double glazing, go to an alternative energy shop, and ask for the plastic double glazing. You attach it like Gladwrap, heat it with a hair drier till smooth — and you get excellent insulation and a lot more peace. It costs about $50 to do an average house. It only lasts a year or two and tears easily — but is simply and cheaply replaced.

### 3. Courtyards and walls

If your house is in a noisy spot, try building a courtyard in front of it to help block the noise and a courtyard on the most peaceful side too, for you to enjoy. A small fountain will also create peaceful 'white noise' that will help block out more unpleasant sounds. Large rocks and mounds in the garden too can help block traffic noise — not as much as you may think, sadly . . . but they do help.

### 4. Plant shrubbery and tall trees

Even a hedge won't block much noise, though it will create 'white noise' — rustling leaves and birds that will hide the problem a bit. But lots of shrubs and tall trees will create a myriad of broken breezes, disturbing the pattern of sound so it isn't as intrusive — and reduce it too, a bit.

## PRIVACY

This is easier than soundproofing of course. It's often difficult to tell exactly where noise is coming from, as it can bounce around. But if you want privacy, just work out where eyes can penetrate — then block them off.

### Fences

Tall fences work well. They don't have to be palings, stone, brick, or aluminium cladding to give privacy. Even netting fences can be planted with evergreen climbers like passionfruit, banana passionfruit, or jasmine, for fast protection.

The taller the fence, the greater the privacy. Tall fences, however, can loom a bit, and make your garden claustrophobic. If your fences are made of stone, brick or concrete, install 'moon gates' or small windows in them . . . you'll have a sense of being able to look out, but unless passersby stop and peer, they won't be able to see in.

### Hedges

Hedges are a haven for birds; also, on the debit side, rat nests and rubbish. They can also die on

**When privacy is important, the space around your house can be broken up by trees and shrubbery.**

you, which fences won't, and will need some maintaining, anything from a trim once a month to every couple of years, depending on what your hedge is made of.

One of the best hedges for frost free — or light frost — areas I know is avocadoes. Plant a metre apart and in five years you'll have dense cover that can be pruned to shape and a heck of lot of fruit. Photinia make possibly the fastest hedge around — not too bad if pruned. Kumquats aren't bad either, though slower growing and camellias can be wonderful.

Tea tree and callistemon, by themselves or together, make wonderful bird attracting hedges if you want a more casual look, as does grevillea rosmarifolia as long as it gets full sun . . . it can look a bit mildewy in some climates.

## Lattice

If you're close to lattice you can see through it but passersby at a distance can't. Lattice is wonderful to place in front of exposed rooms like laundries, or partially enclosed verandahs.

## Shrubbery (I love that word)

Most passersby aren't Peeping Toms, just casual glancers, and if the space around your house is broken up by trees and shrubbery they won't crane to get a closer look.

## Curved paths

*Ditto.* People mostly look in straight lines and a curve will divert their glance from your front door or garage till they're well past.

## Courtyards

This is not a project that you would expect to knock over in a weekend but on a busy road, consider one. Or even on a not so busy road a courtyard is a marvellous retreat, a sanctuary from the world, a small, controlled space easily maintained and peaceful. Just add water — a pond or a trickling fountain. A stone walled, stone paved courtyard next to a stone house is magic . . . or even if your house isn't made of stone.

## Terracing

Terracing makes a small space look bigger. You'll feel you have a lot more *lebensraum*, even if you haven't. Terracing also gives more opportunities to plant different levels of trees and shrubs, and this multilayered greenery will be more effective blocking both noise and prying eyes than a single level would be.

## Windows

Well, have curtains, of course — that's why the 50s were so hot on lace curtains permanently across windows as well as other curtains behind them. Those inside could see out, but no one could see in.

Venetian blinds do the same thing — and so do wooden shutters, but much more beautifully — and they also cut out some noise (see above) and let you adjust light and glare as well. And tinted glass.

## Pergolas

In this house we see the world through leaves — kiwi fruit leaves out some windows, grape leaves out others, rose leaves, passionfruit leaves, honeysuckle leaves, jasmine leaves. There isn't a window in the house that has a view that's quite unclouded by greenery.

This has its advantages. The only disadvantage is autumn mess — leaves that actually aren't a disadvantage because they get turned into mulch and new gardens; and slimy kiwi fruit flowers underfoot in November; though I suppose it's a bit Scrooge-like to complain of excess flowers.

Firstly the house is cooled — more shade, more moisture and in winter the leaves fall and we get sun instead. Secondly, rustling leaves, the 'white noise' mentioned above. Thirdly, the fruit — more than we can eat. Fourthly, the birds, gaily consuming the rest of the produce, more entertaining than any TV as they argue and contort over the most succulent of the kiwi fruit. And fifthly . . . well damn it, life's better seen through greenery. What more can I say?

Building a pergola is easier than you might think. All you need are poles, wire and a spade. The poles can be treated or not — treated lasts longer. Recycling centres often have cheaper stuff; but check that it's solid; and if you are VERY canny you can make a mould and make them out of concrete: cheaper, longer lasting but not quite the same gentle rustic look.

First, mark out the site with string; also mark out where each post will go, and make sure your top pieces will stretch from pole to pole with a bit left over.

Dig a hole — at least 600 mm deep. If you are being very professional, add four cups of quick drying concrete to each hole to make a firm footing for the poles to rest on. (I must admit, 30 years and many pergolas later, I have never done this! But professionals advise it).

Place the post in an upright fashion; if possible use a special level to make sure it is vertical! Replace the soil or fill the hole with quick drying concrete.

Hold the post in place while the concrete sets — about five minutes. (This is why I use dirt!) In termite-ridden areas fill the hole with concrete instead and insert a metal stirrup to hold the pole or use Pole Saver Rods, a non-toxic strip of boron and fluoride that diffuses through the wood and can prevent and destroy termite attacks for up to 10 years. (Contact Green Harvest (07) 5494 4676). Ask your friendly hardware person for items and instructions.

**Building a pergola is easier than you might think. All you need are poles, wire and a spade.**

Now measure the height of your first pole and make sure all the others will be the same height. It's easier to cut them on the ground than standing up! Place the poles across the top and wire them onto the uprights or bolt or nail them in place.

Then plant, water, feed and wait.

Most climbers will give you some fruit after a year (faster than fruit trees) and a lot of fruit after five years or so. As for which you choose it's up to you and your climate. Grapes grow anywhere (ask your local nursery for the most suitable varieties), *ditto* roses, kiwi fruit anywhere south of Brisbane, hops *ditto* (hops shrivel up to nothing in winter), various passionfruit varieties depending on where you are. The most cold tolerant are banana passionfruit — great pink flowers and banana-shaped fruit: not very sweet unless they get warmth but good with sugar added on a pavlova or in cordial or ice cream, and better than no passionfruit at all. We have them

**Val Plumwood's stone house in the rainforest shows great care and artistry in the laying of stone.**

growing over our chookhouse.

And underneath the pergola you pave a terrace for dining and wining and dalliance or roller blading, warm in winter, cool in summer; the most boring house possible looks infinitely better with a pergola or two or three. If I were to do one thing to a garden to improve the quality of life I'd build a pergola.

**Note:** allow a whole weekend to build your pergola and plan it the weekend before. Ask a couple of friends over to help too, with food and grog but don't drink the grog till afterwards or the poles won't be straight. Two of the poles in our large shed still have a definite lean due to not locking up the beer till afterwards.

*It always infuriates me when people say how lucky we are to live in this house, in this garden! Luck had much less to do with it than callouses on the fingers, aching backs and determination. It WAS hard work. It did fill up large parts of several years, on and off over the past few decades. But I wouldn't have missed the experience, and I would never ever like to live in a house I hadn't built or at least had a hand in building and designing, where I wasn't connected to every lump and window by a hundred memories.*

SIX

# Rocks in the Garden*

I LOVE rocks. In fact you have probably guessed that by now. I passionately, absolutely adore rocks. Giant, gorgeous, ornamental rocks instead of garden gnomes or statues of kangaroos throughout the garden; smaller rounded rocks outlining gardens; flat rocks for paving and wider flat rocks for garden stairs, longish rocks for garden walls . . .

My love of rocks probably explains why my back is in the condition it is today, but that's another story (well, actually, it <u>is</u> this story).

Twenty six years ago the area that is now my garden was a mess of weeds and rocks. I used the weeds to make compost so now we have decent soil instead of clay and the rocks have become all sorts of things.

Bryan also adores rocks. Give him a dozen good-looking rocks and he'll turn them into a set of rock chairs (a bit cold on the bum but nothing that a cushion won't fix) or another rock path.

Bryan once spent six months, on and off, crowbarring out a giant, flat topped granite boulder and hauling and winching it about 200 metres plus jacking it up and poking smaller rocks under it to make it level to become our barbecue table. We are probably the only people outside the Flintstones to have a giant, solid, granite boulder table.

Our garden steps are made of rocks. Our garden beds are edged with rocks. We have rock gardens made of — you guessed it, plus a rock edged pond and when we go swimming in summer it's in a rock pool in the creek.

Why rocks? Well, according to classic Feng Shui theory, and me too, there's nothing like a rock for centring a garden. A really large rock, or group of large rocks, really does give you a sense of solidity, safeness, being centred on the earth — all airy fairy terms but, believe me, there is something very comforting about a good-sized rock nearby, all curves and gentle angles.

Rocks also come in an infinite variety that manufactured paving stones, bricks *et al* can never have. Every time I walk up our stone entrance stairs I notice another fascinating angle or bit of colouring (not to mention a few more weeds between the crevices).

One of the joys of having real plants instead of plastic ones is that they do vary. And it's the same with rocks. No two rocks are the same. The whole point of a garden is that it ties your everyday world into the natural world of plants and animals . . . and rocks.

## Where to get garden rocks

See chapter two. Also, look in the *Yellow Pages* in the phone book under 'Landscape Suppliers'. Ring a few places and ask them how much their rocks cost, how much it will cost to deliver them and what sort of rocks they are. A quick question about where the rocks come from is in order to make sure that the environment is not being pillaged — bush rock removal, especially in the Sydney sandstone area, has driven some species, especially of reptiles, to the brink of extinction. Be prepared to pay more for rocks from a reputable and responsible supplier.

The cost will vary enormously, depending on where the nearest quarry is. Sometimes you can get 'seconds' rocks from quarries for about $45 a tonne; sometimes you'll be paying several hundred dollars a tonne. But, on the other hand, rocks don't need feeding, watering or pruning — and, unlike other aspects of your garden, no matter how badly you treat them they'll never ever die!

## Artificial rocks

A few years ago if someone had mentioned artificial rocks I'd have snorted. Artificial rocks? You may as well put a plaster of Paris palm tree in the

*Warning: all this advice is suitable for average backyards — not areas of heavy use like parks or kid's playgrounds, where everything must be much more sturdy!!

**Bryan once spent six months, on and off, crowbarring out a giant, flat topped granite boulder and hauling and winching it about 200 metres to make our barbecue table.**

front yard . . .

Then I saw an exhibition of quite stunning water features at a garden expo. It was only after about half an hour of wondering how on earth they'd managed to drill so neatly up into the rock that I realised I wasn't looking at 'real' rock at all.

Some artificial rock looks exactly that — artificial. Others look wonderful — and waterfalls and ponds made out of them are much cheaper, lighter and more convenient than the real thing as well as avoiding the ethical problems of the rock's removal from its original position. Again, look in the *Yellow Pages* for suppliers.

## What to do with rocks

### Rock edged gardens

Rocks around a garden help stop weeds invading, not to mention small children and their bicycles. You can also admire your rocks even when the plants aren't looking their best.

If you are going to edge your garden with rocks, ALWAYS LAY THE ROCKS BIGGEST SIDE DOWNWARDS. I know it's tempting to lay them on their sides so they look bigger — but somehow rocks always look a bit silly like this, like giant teeth grinning up at the sky. Rocks around gardens need to be large, and they also need to sit properly. Rocks ALWAYS need to be placed on the ground the way they would have laid before being removed.

It's a good idea to tread on the rocks you place

### Preserving lichen on rocks

*If you are moving rocks in the garden, always wash the dirt off them as soon as you've got them in place. Dirt and dust can kill moss or lichen. Remember that if you are moving the rocks to a drier spot, or one in full sun, their moss or lichen may die. The extra humidity around a small pool or swampy area on the other hand, may really help lichen and moss to grow.*

## How to move an incredibly large rock

*I am not exactly advocating moving large rocks, even though we've done it and will keep doing it. Moving large rocks is dangerous. They can roll on top of you, crush your fingers (I no longer play the violin due to an accident with a large gas bottle rolling back . . . just thought I'd better be honest here) and unless you are very careful you can injure yourself in a whole variety of other ways. Move rocks at your own risk!*

*If you do decide you want to move some whoppers, then you'll need:*

- *a crowbar, or better still two or three crowbars*
- *sturdy bits of wood (100 mm X 50 mm)*
- *other small but strong rocks of various sizes*
- *possibly a pulley*
- *patience.*

*Moving large rocks is a learnt skill. You learn how to wriggle it up a bit with a crowbar, push a bit of wood or another rock in the space, then wriggle it up a bit more. You learn to use bits of wood or rocks as fulcrums with your crowbar as a lever. One of our crowbars is decidedly bent due to this.*

*Don't rush moving large rocks. We have a giant flat-topped rock we use as a garden table. It took Bryan six months, on and off, to move it to its present spot. Every day he'd wriggle it a bit more, study the angles, and work out exactly how he was going to tackle the next bit before he did it.*

*It seems almost impossible now that one man ever moved a rock so large so far. But he did, simply because he was patient, persistent and methodical.*

*If you want to move very large rocks from place to place, you can:*

- *Invest in or hire a sturdy trolley.*
- *Lever them up a plank ramp onto a couple of logs, and keep the logs slowly but steadily rolling at the same rate, so the rock doesn't fall between them. It inevitably will at some stage if you're taking them very far. Then you'll have to lever them up again. This method is only suitable for flattish ground!*
- *Attach a pulley to a tree (always use a tree protector — see your nearest 4WD shop) or other solid object, and haul them that way. You may find there is no tree in exactly the right spot to attach the pulley to. In this case the shortest distance to the spot you're aiming for may not be a straight line! You'll have to pull the rock to another spot where it can be pulled from another direction . . . by the time you've finished the rock may have zig-zagged several times!*
- *Use planks as a ramp to wriggle them up onto a truck. It's a good idea to attach a pulley or other safety chains to stop them rolling back.*

around the garden, before you leave them there permanently. If they don't stay firm when you step on them, bed them down more deeply into the soil, or try laying them on another face. All rocks need to be . . . well, rock steady. Even if you don't intend jumping on them in the future, you can be sure some kid will.

### Single great big rocks

Why bother with a statue of a nude Greek god when you can have a rock? One great, gorgeous, massive thing, embedded in the soil the way it would be naturally: don't try to make it look taller by sticking it at an unnatural angle — the biggest side must go downwards. When in doubt, choose a round rock!

If the rock has been fairly freshly quarried you can often see the change in colour that will indicate where it was previously embedded in the soil — you will always get it right if you follow this and bury the rock exactly as it was in nature.

Or have a group of rocks — one great big beauty, with smaller ones around it to make a rockery.

## Garden paving with rocks

See 'Compacted paving floors' on page 35. You can use exactly the same techniques in the garden, though in the garden absolute flatness isn't as essential, unless it's an area that's walked on a lot — or will be used for garden furniture!

Stone paving can be rough and rustic-looking using reasonably flat rocks firmly bedded into sand, instead of concrete and mortar. You can also buy sawn stone for a more elegant and perfectly even effect. Again, consult the *Yellow Pages*. A range of much less expensive synthetic stone pavers is also available in a range of colours and textures, but do make sure that any synthetic stone will tolerate a fair bit of wear and doesn't stain if stuff is dropped on it — the quality can vary.

If you are paving an area that will be used a lot — either walked on or with furniture on it, it needs to be both flat and level. If it's a little-used area, you can use more irregular rocks.

All paving must slope away from your house. If your house is on a slope and you are paving uphill from the house, make sure it slopes on either side of the house, and that the water won't back up at any point and flood the house.

If the paving is going to come right up to your house, make sure that it's well below your house's damp course. Ideally paving should be about the same height as the ground around it. That way you don't stub your toe . . . and your house will also be safer from damp transmitted from the paving or the concrete it's on.

Remember that paving will increase run off — do make sure you know where that run off will end up!

## Stone in sand

It's NOT a good idea to set stone directly into the soil — it can't bed itself down firmly enough and it is very hard indeed to get it level. It's best to set them in sand instead.

Rocks in sand should be reasonably large, as tiny ones are more likely to dislodge, even if they are between the larger ones. This doesn't matter if they are going to be laid in mortar.

1. Clean away all grass, rocks, tree and grass roots.
2. Take a good hard look at the slope of the land. ALL paving should slope away from the house. The slope can be very slight if you are going to place furniture on it, but if you are in an area that gets the occasional deluge — and I don't know any spot that doesn't — you must have some slope.
3. Put down a layer of sand. This needs to be about 25 mm thick, but if your rocks are thicker than that or bits of them are thicker if you are using irregular rocks, your bed of sand must be deeper too.

   The simplest way to lay down a bed of sand is to take lengths of wood the same length as the bit you are going to pave and at least 25 mm high. Keep them in place with small pegs of wood on either side.

   Pile the sand into this framework, rake it reasonably level, then level it off with another long piece of wood. When I say 'level' though, I don't really mean level — remember that your paving needs to have a slight slope, so your bed of sand does too.
4. Place stones on the bed of sand, starting at the corners and perimeter and working inwards. Tap each one down into the sand with a rubber mallet, or wriggle gently into place.
5. When the whole lot is finished, sprinkle more dry clean sand onto the paving and sweep it over and over the paving, in all directions, till all the cracks are full.
6. Wait a few days, then sweep in more sand, as the first lot packs down.

## Plants in paving

Unless you mortar your paving together, weeds will grow in the cracks. These can be killed with a kettle of near boiling water, or watered and pulled out but the easiest way to control weeds is to plant something else.

## Suitable plants for crevices in paving

**Alyssum** — this is great for hot dryish paving, and reseeds itself readily.

**Baby's Tears** — great for shady damp areas.

***Ajuga reptans*** — suitable for damp shade or damp sun.

**Thyme** — creeping thyme or everyday clumping thyme is suitable, but as thyme gets woody after

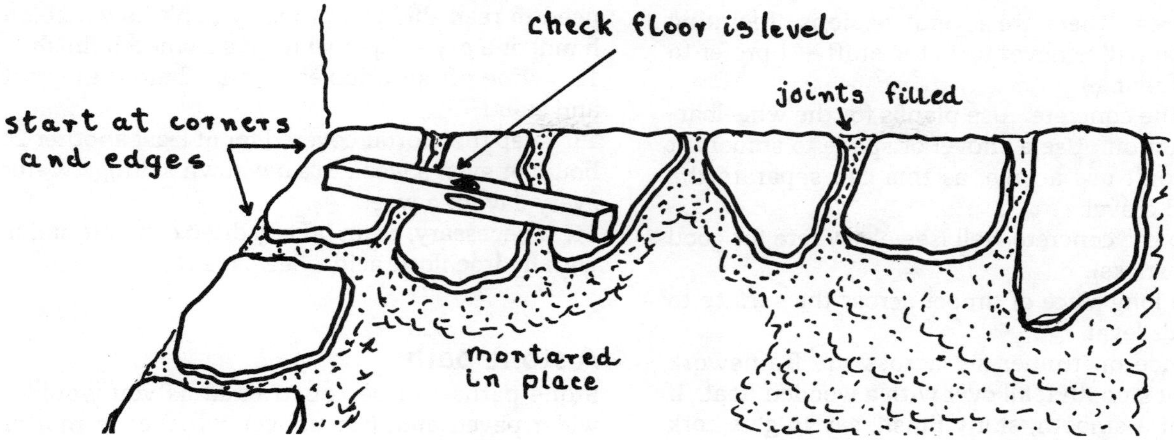


**LAYING ROCK PAVING**

three to five years it will need to be replaced. In the garden you can cover the lower stems with soil so they grow new roots and rejuvenate the plant, but this is difficult in paving. Thyme needs full sun. Thyme can look wonderful though if you deliberately leave spaces for it.

**Chamomile** — needs full sun, moist soil; use flowers medicinally and for tea.

***Pennyroyal*** — needs damp soil and semi but not deep shade. The hotter the climate the more shade it tolerates.

**Dichondra** — good for damp, shady paving.

**Corsican mint** — tiny leafed, fresh smelling mint; looks stunning and smells extraordinary when trodden on; won't take much traffic though and must be kept moist. It needs full sun in cooler areas, semi-shade in warmer climates.

## Laying rock paving in concrete and mortar

If your stone is less than 50 mm thick (as split stone and slate often is) then you will need to lay it on a bed of concrete and mortar or it will crack.

If you are laying a driveway, you'll need a bed of concrete about 100 mm deep. Even if no vehicles will be driven over it, it should be at least 75 mm deep. Remember that about 20 to 30 mm of mortar and rock will go on top of this, so you may have to dig deeper than you first think to ensure that the top of your paving is at ground level, not above.

Colouring can be added to all concrete and mortar, and stone paving looks far better with a little bit of colour added to the mortar so you don't have dead white or grey between your lovely rocks.

1. Prepare the site as for laying stone in sand.
2. Place a framework of timber around the levelled site with wooden pegs every metre or so to keep it in place. Paint the framework with old sump oil (use gloves and goggles!) or linseed oil to stop the concrete sticking to it.
3. Lay down plastic if your paving is indoors.
4. If the paving is indoors or on an unstable site or will have heavy traffic over it, you need reinforcing. This helps stop slabs cracking. Most outdoor paving though won't need reinforcing, so skip this bit.

Lay the reinforcing on 'chairs' so it's in the top half of the slab, but will still have 25 mm of concrete above it. Reinforcing shouldn't be within 50 mm of the edges. Allow 250 mm of overlap when you join two bits of reinforcing together.

Reinforcing can be cut with bolt cutters or a hacksaw — bolt cutters are easier. It is even easier to work out the sizes you need and have them cut where you buy the stuff.

There are different thicknesses of reinforcing, depending on how thick your slab will be. Check the thickness you need with your local building inspector.

5. If you don't use reinforcing mesh, you may need to add expansion joints. Concrete expands and shrinks in heat and cold, and if there are no expansion joints it will crack.

The easiest way I know (and remember I am not an expert) is to leave a bit of wood every six metres and next to any solid wall, like a house wall. Hoik it out when the concrete sets, then fill the space with 'jointing compound' available at hard-

ware stores. There are several versions of jointing compound. (I've never used the stuff — I prefer to use reinforcing.)
6. Pour the concrete. Use planks for the wheelbarrow to run on. Use a shovel or spade to smooth it out — don't use a rake, as this can separate the sand and gravel.
7. Tamp the concrete well (see elsewhere for good tamping sticks).
8. Pull a long piece of timber across the surface to ensure it's level.
9. Balance on timber set across the framework while you smooth it all over with a wooden float. If you want a smooth, shiny finish, say to glue cork tiles onto or to polish and live with, you'd float again with a metal float.
10. Cover it for a week to stop it drying too fast and cracking. Water it after about 24 hours and then water every day for a week.
Note: concrete is dry enough to walk on cautiously after about half to one hour in summer, and 10 to 12 hours in winter.

## You are now ready to add the stone and mortar!

1. Leave the boxing in place till after the mortaring and rock-work are completed.
2. Lay your stone out on the slab in the way it will eventually be mortared in. You need to work fast with mortar, so everything needs to be in place.
3. Hose the concrete and stones for about five minutes to let the water really soak in.
4. Mix the mortar with 3 parts sand to 1 part cement, with just enough water to hold it all together. If it's too wet it'll all squish up above the level of your rocks, and they'll tend to tilt down into it too.
5. Start at the corners then work inwards. Pick up one of the rocks; slap down about 20 to 30 mm of mortar, replace the rock in the mortar, and tap it into place with a rubber mallet. If the tap sounds hollow, it's not bedded properly.
6. Wipe off all concrete stains as you go.
7. Keep scraping across the site with a long piece of timber to keep it all flat and 'level' (slightly sloping).
8. When it's all done cover it with sacking, a tarp or black plastic to stop it drying too fast.
9. Leave it for 24 hours to set.
10. Water lightly.
11. Take more mortar (one cement to three sand) but make it a little wetter this time . . . not runny, just oozing. Use this to fill the cracks between the stones. Use a bricklayer's brush (though if you need to read this you probably don't have one on hand) or a dryish sponge to give a smooth finish.
12. Wipe off all dribbles at once before they set and stain!
13. Keep the mortar covered for at least another 24 hours or even a week to slow down setting. Water every day for a week.
14. If necessary, clean off any dried concrete stains with hydrochloric acid. See page 70.

## A stone path

Stone paths can be constructed as you would a wider paved area (see above), whether in mortar and concrete or in a bed of sand. Stone paths should have larger rocks on the outside; this makes them more stable to walk on, and gives them a more stable look too.

Don't ever have a row of rocks going from one side of the path to the other, or in line along the path. Keep the appearance random.

Whenever you make a path — stone or otherwise:

- make sure the path goes as directly as possible from point A to point B . . . otherwise people will ignore the path and go by the most direct route!
- curve the path as much as you can, keeping the advice above in mind . . . straight paths look boring and rigid. Straight paths also look tatty unless the grass on each side is trimmed; you can get away with a lot more mess with a curved informal path. Curved paths also help a feeling of privacy. If there is a straight path leading up to your front door, the gaze of the onlooker will be drawn to your front door. A curved path fools the eye, at least temporarily, which is all that's needed if the onlooker is speeding past in a car.
- watch out for stubbed toes! Keep the top of any path as smooth as possible; rock edges (or paving or brick edges) should no protrude. The exception to this are very rounded smooth rocks . . . but even then, I'd prefer a flat path!
- remember that people on crutches, wheelchairs, strollers, prams and on bicycles may use your path (you may well be one of these people; well, okay, not the pram perhaps).

## Rock steps

Rock steps can either be lovely irregular affairs, mimicking the way the land might rise naturally as though the rock just happens to be in exactly the right place for your garden. You can also use split

**Building Stone Steps**

blocks of sandstone or other rocks for neater, more elegant steps. The effect you choose depends, of course, on the nature of your garden. Manicured lawns and neat garden beds need elegant steps to complement them; irregular natural stone steps fit best into a rambling native or cottage style garden.

## How to build stone steps

First of all, consider if you really want steps!

The day may come when you or your friends can no longer manage steps. Steps are also difficult to negotiate with armfuls of firewood, gas bottles and other awkward items. If possible, construct ramps rather than steps, and make the slope of the ramp as gradual as you can. Ramps can be paved with stone (see 'Laying rock paving' on page 59) or made of concrete or simply covered with grass or other ground covers like dichondra.

Steps are made up of 'treads' — the level part of a step and 'risers' — the bit that goes upwards. Most steps should be at least 300 mm wide unless it's only going to be used by small-footed elves or pixies, with a maximum rise of 180 mm. All steps should rise at the same rate. If they vary, inevitably someone will trip.

Work out the plan of the steps before you begin building them. Divide the height of the total steps by 180 mm (or less) to work out the number of steps you'll need. Then multiply that by 300 mm to work out how far your set of steps needs to extend.

Round the resultant numbers up or down to arrive at a reasonable compromise number for your flight of steps keeping in mind the provisos above — many people find that a quick sketch with some attention to scale is very useful at this point.

Don't feel you need to make the outside edges of your set of stone steps straight and vertical. You can let them slope away, or edge them into the surrounding soil.

Stone for steps must have reasonably sharp edges. If they are too rounded it's hard to know exactly where to put your foot, and that can be dangerous. Make sure the steps slope very slightly — VERY slightly – downwards so that water can run off. This is especially important in cold areas where any water might turn to slippery ice.

## Free-standing steps of stone and concrete

1. Build the bottom step. This will be very wide — wide enough for all the 'treads' of the rest of the steps to fit on it. Follow all the suggestions in 'Laying rock paving in concrete and mortar' on page 59

for foundations, drainage, framework and mixing.
2. Proceed upwards. Make sure no bits protrude or slope too much, and that every step is about the same. You need to be able to walk up steps without thinking about it!

## Stone steps in dirt banks

Our garden is on a slope, in fact there is nowhere really horizontal on our whole property. Our garden steps are all dug into the slope instead of plonked on their own foundations. While in my youth I bounded up slopes like a mountain goat, nowadays I prefer to proceed in an orderly fashion, up steps.

1. Start with the bottom step. Remember that the total rise (the stone plus the dirt below it) should be no more than 180 mm. Make sure it is flat, or at least flattish on top, and not too rounded, or it will feel insecure when you tread on it. Also make sure it is firmly embedded in the dirt, no matter how hard you jump on it.
2. Dig a flat platform for the next step, again with no more than 180 mm of rise from the lower one, and at least 300 mm back from it. The narrower your steps, the deeper the treads should be. Your steps will probably have to be several stones long and wide to be comfortable. Again, do the jump test for each step. If it's not firm enough to jump on, it's not firm enough!

## Rockeries

Rockeries are mounded gardens of rock and soil. They give you a lot more garden per metre and are a great way of making flat gardens interesting or of irregularly terracing sloping gardens.

As they are sloping they are also well drained and help plants catch the light if they are angled to maximise the sun — great for sun lovers like lavenders, prostrate rosemary, miniature convolvulus, early tomatoes or eggplant. (I use our rockery for cold sensitive perennial chilli, heliotrope, etc. They wouldn't survive without it!) You can also create pockets of moist shadows which will suit ferns and mosses.

## How to build a rockery on a slope

1. Lay down your largest rocks for the bottom layer, in a rough semi-circle. The rocks on the bottom layer should be touching each other, and the largest rock needs to be in the middle with the next biggest rock next to it and the smallest at the far end. Fill the circle or semi-circle with easily draining soil (NOT clay) and stamp down well.
2. Build the second layer a little further in from the first layer. This layer can be more irregular, but again the largest rock needs to be in the middle. Make sure the gaps between rocks on the bottom layer are covered by rocks in the second layer. This interlocking will make the wall stronger. Back fill with soil.
3. Add more layers according to the height of your slope and the number of available rocks. Each layer needs to be both higher up the slope and slightly inwards, with the largest stone in the centre. Rocks should get smaller the higher up you go otherwise the rockery will forever look as though it will topple and fall — even if it doesn't.

Make sure that every rock is firm before you add another layer. You should be able to stand on each rock without it rocking.

## Freestone (not concreted or mortared together) rock retaining walls — for example, against a hill

Free stone rock retaining walls look great and are surprisingly quick and easy to build yourself — much easier than a free-standing rock wall and much, much easier than a mortared rock wall.

1. Buy or forage rocks that are no larger than you can comfortably handle, but still as big as possible — some rock types are much lighter than others, so ask what suppliers have in stock. If you are buying them, have the rocks dumped as near the site as possible.
2. Make sure the slope where you want to build your wall slopes back by about 5 to 10 degrees — a gentle backwards slope. This makes the wall more stable. You may need to dig it so it slopes.
3. Remove any grass or weeds so there is a firm footing. You don't have to build foundations for a free stone retaining wall as any slight movement will be absorbed by the wall.
4. Choose the largest rocks and lay them along the slope, then 'backfill' with dirt and stamp it down well. In very wet areas, or if your soil is likely to be damp, install an agricultural drain at the bottom of the backfill leading out from both sides of the wall. You can buy agricultural drain material at a hardware store. In very wet ares — or areas that have periodic deluges — it's best to 'backfill' with rocks and gravel, to improve drainage.
5. Lay the next row of rocks, making sure the gaps between the rocks on the bottom layer are covered by rocks on the second layer. Backfill again. This interlocking will make the wall stronger.
6. Continue row by row, back filling and stamping firmly.
7. Level off the soil behind the wall, plant the top

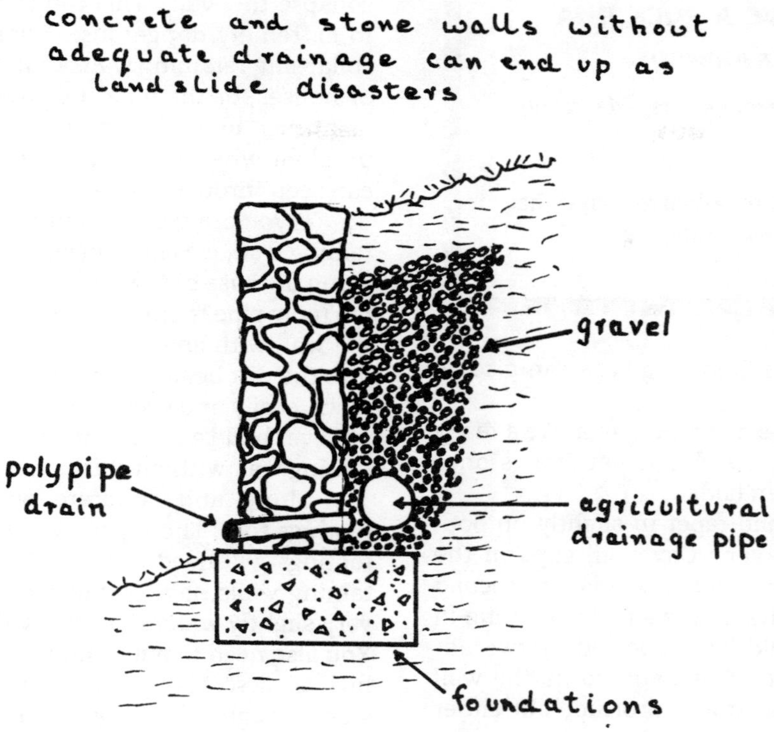


**Cross Section of Stone and Concrete Retaining Wall.**

with ground covers and plant alyssum or other small creepers in any soil-filled crevice in your wall.

## A formwork stone garden wall

Freestone garden walls are lovely — but you need far more skill to build them than you do with formwork. Freestone garden walls have to hold themselves up. In a formwork wall the concrete sticks it all together. A formwork wall built with skill looks better, but one built by an absolute beginner will still look good — and stay up! If you want to build a formwork wall in the garden, again first dig your footings. Because the wall won't bear any load except its own weight about 30 cm deep of concrete is enough. As with house walls, make the footings slightly wider than your walls. Then proceed exactly as you would for a wall for a house except do make sure you top the final piece of wall with good looking rocks and smooth it off so it's relatively even.

Stone garden walls are great sun traps. If you want to grow lemons in a cold climate or have tomatoes fruit in late spring, you need a few north facing stone walls. The stone will reflect sunlight onto your plants during the day, and at night they'll retain warmth, and warm your plants all morning. In very cold climates 'walled gardens' were the only way of growing enough vegies in the cold and cool months to feed you or for having fruit and berries ripening early too. And they are an excellent way of making a small courtyard a refuge from the outside world.

## A freestone rock wall (without formwork)

I have only ever built one freestone wall! But at least it's still standing.

1. Choose your rocks. They'll need to have at least two flattish sides, a bit like irregular bricks. In fact the more they look like giant bricks the better! It is very difficult (though not quite impossible) to build a freestone wall with round rocks!
2. Make sure your soil is solid enough for a stone wall: no fill or slippage.
3. Clear away any grass, and mark out where the wall will be with pegs and strings between them.
4. Dig out a 300 mm deep foundation. This foundation need not be concrete — in fact it usually isn't — just a layer of gravel (see below) and stones. The width will depend on the height of your stone wall, but should be AT LEAST 600 mm to be stable. If your wall is going to be more than 900 mm high, the base should be two thirds the height of the wall, so if you are planning a 1800 mm wall, the base should be 1200 mm wide.
5. Lay down about 200 mm strong gravel or road

### HOW TO MOVE A ROCK IN A WHEELBARROW

1. *Lay the wheelbarrow on its side facing the rock.*
2. *Roll the rock into the barrow.*
3. *Lift and roll the wheelbarrow upright with one person on each side.*

base. A concrete and reinforcing base can also be used.

6. Lay the largest stones down first, to make a really solid base. Fill any gaps with smaller stones once the big stones have been laid.

7. The next layer should taper in slightly on both sides from the first layer. Cover all gaps in the lower layer with rocks: in other words, the second layer should overlap the first, each major stone in the second layer should be supported by two adjoining stones below it. This helps to tie the wall together. Make sure no stones protrude out either side and that none of them slope down, even slightly, or you may find the stones on top slide off.

If you've never built a stone wall, you may find it helpful to make a sloping wooden 'guide' to make sure you achieve the correct slope so it's nicely flat on the bottom and skinnies on top. The best slope is about 1 in 10 to 1 in 25.

If, like me, you have no idea what this looks like in practice, get three bits of wood, one as high as your wall will be, one about half as long again and a shorter piece. You'll also need a couple of cheap plastic set squares (from the newsagent). Arrange the bits of wood on the ground. Attach the short bit of wood to the bit of wood the same height as your wall at an angle of 90 degrees (check with set square).

Now take the third bit of wood and join it at an angle of 85 degrees to the longer bit of wood. Join it up to the shorter bit at the top; saw off what isn't needed, and use that to guide you.

7. When you reach the top of the wall, cap it with large level stones. If your wall is on a slope, place poly pipe or other drains through the base of it in case water builds up on the uphill side and pushes over your laboriously-built wall. It's also a good idea to dig a large drain, filled with rubble or agricultural drain pipe, on the uphill side.

## Stone and concrete retaining walls

See the diagram on page 63

Stone and concrete retaining walls can have major — even disastrous — problems. If a wall is holding back soil, then moisture can build up and crack or collapse the wall. This can range from a nuisance to extremely dangerous. There have been many collapsing retaining walls that have crushed people or houses, or the build-up suddenly collapses in a great rush of watery mud. This isn't as much of a problem when the wall isn't concreted, as water can seep through the gaps.

If you are going to build a formwork stone retaining wall, leave drainage holes every 30 cm or so along the base of the wall, and at a few points above too (poly pipe is great for this). Fill the space behind the wall with about 30 cm of coarse gravel too, before the dirt begins, so that any moisture can drain down easily and escape out the drainage holes. Do not ever build a stone and concrete — or brick — retaining wall without this!

Stone and concrete walls will also need footings too. The higher the wall, the deeper your footings will need to be. Our waist high stone retaining walls have 200 mm deep footings, and are very slightly wider than the walls — about 300 mm. You also need to put reinforcing mesh in, on 50 mm high 'chairs'. Make sure it's at least 50 mm from the sides though. If it's any closer, any chips or erosion may mean it's exposed to air and rain, and will rust — and you'll have holes in your foundation, not mesh!

Wait at least 24 hours for the foundations to set before you build the wall. If your wall is going to be more than waist high, or will have a lot of pressure behind it, or if you live in a wet or fierce storm prone area, consult an engineer or at least your local building inspector.

I know this sounds fussy . . . but there are a heck of a lot of tragedies every year with retaining walls collapsing and the whole lot oozing down onto a house. It's best to be absolutely safe.

Once you have your footings, place your formwork up as described in the house building section — balance them on the footings, and tie together with wire, with spacers between. Then proceed as you would for a house wall. The only difference will be this time you will place lengths of poly pipe every 30 cm or so along the wall, and every 30 cm or so up the wall. Make sure each piece of poly pipe extends right to the formwork.

The wall can be one or two or more forms high, as long as you make sure the footings are adequate, and the drainage is too. Pardon the repetition . . . but a wall larger than waist high does need professional assessment to make sure the design and situation is safe!

Place flat rocks along the top of the wall, for a good looking finish. I prefer a slightly rounded edged look rather than sharp edges, on a wall. You can do this by mounding the top layer of stones and

**COMMON PAVING PATTERNS**

concrete up slightly higher than the formwork, so it bulges up in a gentle curve.

When you take off the formwork, DO make sure that there is no concrete in the poly pipe drainage holes!

**PS** And if a retaining wall ever looks a little bulgy, or moves in anyway, run, don't walk, to your local building inspector, and ask for urgent advice!

## An idiot's guide to concrete pavers

Paving is one of the most satisfying things in the garden. Once it's in it's THERE — even floods, bushfires, plagues of locusts, and teenage parties usually leave paving relatively untouched. Paving also helps heat and cool the house (see page 48). You don't have to mow it, chairs and tables sit on it without rocking, and you can see snakes before they see you. It's also good for bikes, tricycles, skateboards and hopscotch.

The easiest way to put down paving is to get someone else to do it — end of story. Unless, of course, you have to take out a second mortgage to pay them. Or you can put your own down, which is a lot easier than you might think.

1. Dig out the site. It needs to be as deep as the paver, plus 50 mm for the sand. Make sure the

ground is level (or levellish — you can still have a gentle slope — but no sudden lumps and bumps). Make sure it slopes gently away from your house, unless you like paddling through your living room.
2. Lay down sand to make a firm flat bed — about 50 mm. Level it.
3. Lay down the paving stones. It's easiest to start along the longest straight edge and wriggle them firmly into the sand — but STAGGER them, so that no edge matches the other. Straight lines of edges look odd. Bash each with a rubber mallet or wriggle them in well.
4. Dust more sand over the pavers and sweep it in — a straw broom is best for this. You can also sweep in a mix of one cement to four sand, then sprinkle with water so they are mortared together and weeds won't grow.

You can fill in any bare bits at the edges with broken pavers, but I just fill them with soil and plant something — much easier than fiddling about with hard-edged fragments. If you do want to cut your pavers to fill in odd shapes, see the instructions for shaping paving stones. I must admit you will get a neater result!

Many paving supply outlets will also give you good advice on laying pavers — and some have demonstrations when you can go and watch the sand bed being prepared and the pavers being laid and firmed.

## How to get your pavers

Look in the *Yellow Pages* — and ring around for the cheapest price and best range. Ask if they deliver too — and how much it costs, unless you really feel up to hiring a trailer and carting your own.

## Making your own

Well, actually it's easy. Even I can do it. I am not what you'd call neat and handy. Buy sand and gravel and cement or bags of ready mixed concrete (very expensive) and a sheet of plywood.

Cut out squares of old ice-cream container — about 7.5 cm high. Add a pinch of colouring to the mix (you can buy it where you buy the concrete). Make sure you measure how much colouring you use so each batch of tiles matches — or at least that the variations are the variations that you had in mind to give a pleasantly rustic look — not a harsh patchwork! I use a mix of 6 shovels of gravel to 3 sand and 1-1/2 cement, with about one teaspoon of colouring, though you may want to use more for a richer colour. You can also of course make much larger paving stones — I just find that ice cream containers make convenient moulds. But if you can nail wood into a rectangle, you can make a mould.

Lay the moulds on the plywood. Fill up the moulds, smooth the tops roughly — wait four or five days for them to harden and there you are. Stack them till you need them.

An hour each weekend for a couple of months or one day's mixing should give you a gardenful — and save you hundreds of dollars. I like to lie a leaf or flower on the plywood before I pour the concrete into the mould — when you lift it off there's a lovely pattern underneath. You can also smooth off the tops then press your hand or foot gently into the mix — and you now have a fossil like record of you or your kids to preserve for the future!

## Where to put paving

- Around the house — then put a pergola over the paving, with a deciduous vine that'll lose its leaves in winter so you get sun through your windows, but it will shade you in summer.
- Instead of the lawn.
- Under the clothes line.
- Create a courtyard for a barbecue or garden chairs.
- Around the vegetable garden and along paths in cold areas — the reflected and retained heat will mean your vegetables grow faster and for longer!
- See also 'Cow Pat' pavers on page 36.

## Tricks with paving

Leave spaces for shrubs — especially if you are using paving instead of lawn. Shrubs will shade the pavers and keep it garden-like. Palms in paving look especially good.

Leave every sixth or so paver out — then plant with prostrate rosemary, alyssum, herbs like thyme, violets — something low maintenance that will happily sprawl and break up the 'too much concrete' effect.

Leave out one giant paver or several small ones; dig a deep hole, line it with pond mat, put in a layer of soil then fill with water and plant waterlilies, Japanese iris, Chinese water chestnuts or add fish to give the cat a thrill . . . and make sure guests don't fall in it.

# SEVEN

# Having Fun With Several Tonnes of Concrete

ONCE you start working with concrete it can become addictive . . . a bit like a grown-up version of plasticine. For those who become hooked . . .

## How to make a homemade pond

Humans evolved in water — and I for one don't feel comfortable away from it — even if it's just a lily pond. Water also attracts birds (you try flying over your suburb and see how much water there is to drink or bathe in — unless you count swimming pools, which birds don't).

Ponds are remarkably easy to make and don't take all weekend either (just a large part of it.) Even if you've never hammered in a nail you can make a pond — it's a bit like mud pies for grown-ups.

**1. Choose the site.** Pools look best if they are where water would naturally drain to: at the bottom of a slope, or at least part way down it, not at the very top of the garden. Mark out the rough dimensions of your pool. If you are going to pave around it, or place rocks around it, mark out that area too. The simplest way to mark out a site is to spread the hose into the shape you want. Irregular shapes look best for informal pools, with curves instead of sharp edges.

**2. Dig the hole.** You can do this by hand, or hire an excavator or bobcat for large ponds that are swimming pool size, or where the soil is very hard and rocky. The hole needs to be at least 50 cm deep if you want to add fish or lilies, so the water doesn't get too hot or cold. Mound up the edges to make the pond deeper. Make sure the edges slope VERY gradually, and try to keep all edges rounded so you don't have sharp corners. Corners need extra reinforcing so should be avoided.

**3. Check there are no rocks** on or near the surface, that may move in time and damage the pond lining. If unsure, dig the hole deeper and line with sieved soil or sand.

**4. Add reinforcing** — this makes the concrete more flexible and less likely to crack. If your pool is going to be less than a metre deep the easiest reinforcing is probably chook wire. If the pool is going to be larger you'll need proper reinforcing, and will need to investigate the whole process in much more detail.

Line the hole with old chook wire (from the dump or redo the chook house) or new chook wire.

**5. Buy sand, gravel and cement** and colouring if you don't want a grey/white finish. Colouring under water may fade, but if you add too much you may weaken the concrete. As a very rough rule of thumb, don't add more than 1 part colouring powder to 12 parts cement powder.

**6. Make a mix** of one part cement to three parts sand and three parts gravel. Add water till it's the consistency of a loose cow pat (or a bit thicker than cake batter for those who have never seen a fresh cow pat up close).

**7. Spread thickly** over the wire in the hole. The layer of concrete will need to be at least 100 mm thick, with the chicken wire pulled up so it's more or less in the middle. There must be at least 50 mm of concrete over the wire. If in doubt, make the layer thicker. Larger ponds may need thicker concrete, as well as heavier reinforcing.

**8. Let it dry a little** then smooth it with a steel float.

**9. Concrete will leach** into your pond for a while, making the water very alkaline. You can stop this by applying a coat of special waterproof paint, or you can just fill the pond with water (it should be waterproof anyway) then let all the water out two weeks later and refill.

This is of course just the basic recipe. Add the odd big rock around the edges or even in the centre. Make sure the seal is good between the rock and the cement. Seal it halfway up the rock if you can. In any case a few, or lots of, rocks around the edges with plants in between greatly increase the attractiveness of the habitat. You can also bung a few rocks in the middle when it's set but gently, gently so it doesn't crack.

There are artificial rocks on the market now which should look disgusting, but don't. If you're

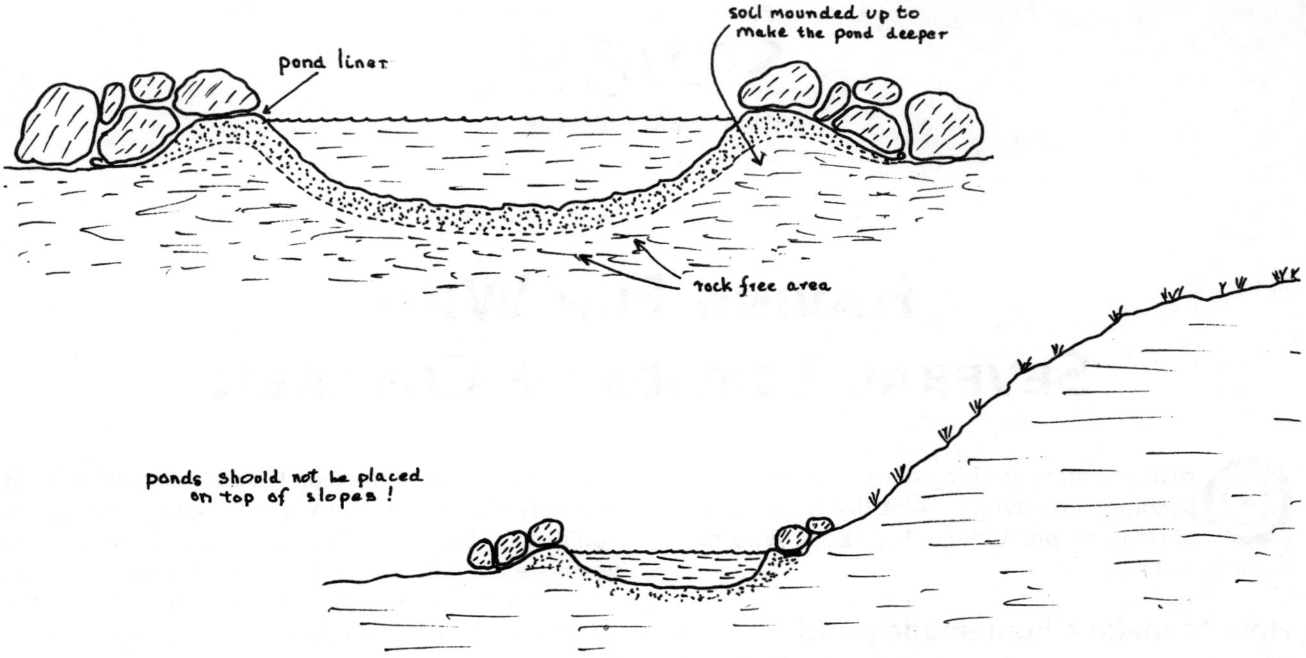


**HOMEMADE POND CROSS-SECTIONS**

prepared to pay enough you can even buy artificial rock waterfalls. The good ones are very hard to tell apart from the real thing and at least you can be sure they haven't been collected from some poor creature's habitat!

Pond liner is even simpler to use than concrete. Probably the best but most expensive pond liner is butyl rubber, which doesn't decay in sunlight and is resistant to punctures. Do make sure though that you buy a brand which won't taint the water if you're going to keep fish.

There are other pond liners that cost less but slowly break down in sunlight. Black plastic is even cheaper but doesn't last and can taint the water too. Don't use it.

Now fill your pond with water. Well-planned lily ponds have a drain at one end so you can let the water out. If you are good with plumbing you'll find several good books in the library that will give you instructions. We just bucket ours out when we want to clean it — the thickish water (what with weed, algae and assorted droppings) is good for the garden. And there are very many buckets full. You can also siphon the water out if you want to water an area below the pond — this is the best method if you don't want to stir up the pond. You can even choose which part of the pond to drain — you can concentrate on the murky layer at the bottom or you can replace the warm deoxygenated water further up.

The easiest way to fill a pond is with a hose. We have a hose connection poking up through the base of ours and an ordinary microjet attached to that. When the microjet is on we have a simple fountain. I say it's ours but the birds probably would claim it too. Dogs are best kept out by saying "No" in a firm voice (assuming your dog is trained enough to recognise the word) or by covering with builder's mesh. Builder's mesh may look ugly but birds love perching on it.

## A Simple Pond

If you want a really simple pond — and have had enough of concrete — go for a pond liner instead.

Now spread the pond liner. There are a variety of pond liners on the market. Plastics are cheaper, but break down on contact with light. The more expensive butyl liners last longer — ask for more advice at the nursery where you buy your liner. You will also need to make sure the pond liner isn't toxic to fish, if you want fish in your pond.

Make sure the pond liner extends at least 100 mm above the water level. Now hide the extra liner with rocks, bricks, stone slabs soil or pebbles, or plant creeping or sprawling plants that will grow right down to the water's edge.

For an even more stunning effect, keep the pool wide and shallow, and cover the whole pond with stones — smaller, round edged ones in the water, (they must be rounded so they look like their edges have been eroded by the water over the past 1,000 years), and larger rocks around the edges.

## How to build an artificial giant boulder

1. **Mound your basic shape from soil**, or soil and

stones, or even from old plastic supermarket bags filled with sand and piled on top of each other. If you think the rock is going to be played on or jumped on (or even MIGHT be played on — kids love clambering on rocks), then it's best to fill supermarket bags with 6 parts sand to 1 part cement. These will gradually set and the structure should be reasonably firm. Jump up and down on this inner material to make sure it really IS firm (cautiously, in case it isn't).

**2. Small 'big' rocks don't need reinforcing.** If your boulder will have more than about 4 square metres surface area, or if the underneath may not be solid, or if you think it may be an 'attractive nuisance' and kids may subject it to rough treatment, then it's best to be safe and reinforce it, either with chook wire as above, for smaller rocks, or with good solid reinforcing. The larger the rock the more reinforcing you need . . . if in doubt, with a very ambitious project, do get an engineer's advice! All reinforcing must be firmly wired together so it doesn't move before the concrete is laid.

**3. Do a few test blocks of concrete** with various colours, to decide what colour rock you prefer and how much colouring will be needed to reach that shade. Remember that you can mix several colours together — and that most large boulders aren't one uniform colour. Study a few natural boulders too!

**4. Spread concrete over the reinforcing**, about 75 to 100 mm thick. Don't smooth it out too much. If you like, sweep over it with a stiff broom in places to give a striated effect, or rake bits of it to give a bit of true life unevenness.

Sometimes metal filings or soil are added to part of the concrete to stain the 'boulder' as they leach out. To be honest most times this just looks like a stained bit of concrete! As your fake boulder ages, moss and lichen and algae will grow on it and stain it naturally, especially if you keep it damp and slightly shaded.

Boulder making is an art. Most look more like concrete lion droppings than boulders but if you do have some artistic talent . . . and a good eye for seeing what the real thing looks like . . . well, good luck!

## How to make very large, wide, shallow pots

Pots are useful things. You can grow frost sensitive trees and shrubs in them, and move them next to a warm sunny house wall in winter. You can use them to create a garden next to the house that can be moved to check for termites and you don't have to bother about invading roots under the foundations and in the pipes.

Fill a wheelbarrow — or even a hole — with damp sand. Make a pot shaped and sized depression. (Another pot is a great way to do this.) Smooth it out. A wide, shallow, pot shape is the easiest to start with. You can get fancier later on as your skill grows.

Mix one part cement, two parts sand and three parts gravel with water until sticky but not sloppy. Add colouring if you want at this stage too. Pour it into the depression. Smooth out the inside roughly (not too heavy handedly or you'll press too hard and ruin the outside shape. The inside shape doesn't matter too much — it'll be full of dirt).

Smooth out the edges and poke a stick down the bottom a few times to leave some holes. Leave it a few days to dry (cover from the rain if necessary). Pull it out and you have a pot. A bit rough and sandy, but this only adds to its rustic appeal. If you don't want a plain grey concrete pot, ochres are quite cheap from hardware stores. You only need a very little to change the concrete's colour. Pots can also be painted, either plain colours or fantastic designs.

## Large square or rectangular pots

### Materials

- one large and one slightly smaller styrofoam box, (the smaller box can be cardboard), or one large and one slightly smaller plastic pot.
- four narrow 'necks' of old soft drink bottles, with sticky tape over the top and bottom so no concrete falls into them.
- sand cement mix — four sand to one cement
- coloured oxides — red green ochre yellow, etc. (from the hardware store)
- water
- a stick

**Step 1.** Add water to the sand and cement mix and stir till it's doughy but not sloppy. Add a LITTLE oxide to change the colour

**Step 2.** Spread 3 cm of the mix into base of the large pot or box.

**Step 3.** Press the soft drink bottle necks firmly into the mix till they touch the base. These will form the drainage holes.

**Step 4.** Place the smaller pot or box into the large one. Pour the rest of the cement mixture between the smaller and larger moulds. Use the stick to poke it in firmly.

**Step 5.** Leave to dry out of the rain for 24 hours.

**Step 6.** Carefully slide or tear away the inner and outer pots or boxes. Remove the sticky tape from the drainage plugs.

**Step 7.** Leave for 4-5 days to 'cure' and get stronger, then fill with potting mix and plants.

# EIGHT

# SIMPLE HOME MAINTENANCE

### How to get rid of grease on the driveway

Warm some kitty litter (unused naturally) in the oven, pour it on the grease, then grind it in with your foot (and shoe). Leave for 24 hours, vacuum up, then repeat as more grease is drawn up to the surface.

### How to remove rust stains from concrete

This is useful when some dingbat has left a rusty drum there — or even a kid's metal toy truck left out in the rain. Mix one part citric acid and one part cream of tartar. Use an old wet toothbrush to scrub the mixture into the stains. Leave overnight, wash off and repeat until stain is faded.

### How to fix leaky guttering

No, it doesn't have to be replaced — not yet anyway. Clean around the hole, lay epoxy filler over it according to the manufacturer's directions (if you're not sure what to buy ask for advice at the hardware section), lay a piece of glass fibre mesh or even fly wire over the hole and press it into the filler. Wait for the filler to dry, then layer more filler on top and under the hole. If necessary sandpaper it and repaint.

### How to clean up mossy paths

I LIKE mossy paths — but they can be slippery. Kill the moss by spraying with one part methylated spirits and one part vinegar. Leave for 15 minutes, spray again, leave till dry and rake or sweep off.

### How to renovate grotty paving

Scrub with baking powder and toothbrush or steel wool. As a last resort dab on hydrogen peroxide (wear gloves). You can also hire high pressure water cleaning equipment.

### How to wash very GREASY barbecues

Try scraping off the grease, then washing with very hot soapy water. If this fails use a mix of caustic soda and water according to the manufacturer's directions.

### How to clean stains off wooden decking

Scrub with laundry bleach — wear gloves and follow directions on the container. This should remove most stains — including algae stains — but you may end up with a 'clean patch' as prominent as the original stain. If the deck is small you may wish to bleach all the wood to make the blotch less prominent. (On a large deck it's easier just to place a chair over the stain.) Make sure you don't kill any plants around too!

### How to remove tyre marks on concrete or paving stones

Hydrochloric acid will wash them off — ask at the hardware store for a preparation containing hydrochloric acid, and use according to manufacturer's directions.

## HOMEMADE PAINTS (AND A FEW OTHER THINGS)

These are really only suitable for concrete, rammed earth, terracotta (great on pots), adobe, concrete blocks, fibro alternatives, stone and other porous materials and not on wood or metal. They also probably won't take if you try to paint them over existing paint.

It's worthwhile experimenting with homemade paint using whatever more or less similar materials you have on hand. The first essential of course is colour — these can be ground ochres, homemade or bought dyes, or — probably the best — colouring materials bought from a hardware store. You get a large amount for about $10.

Secondly you need a 'binder' — glues from fish or animal skins, blood or, most usefully from my point of view, milk, either fresh or dried.

## Basic exterior paint recipe

1 part cement
1 part dried milk
ochres for colouring.

Mix the colours. Don't add too much colouring: keep trying it out adding small amounts of colouring till it seems about right. Don't add too much water or the paint will be too thin and you'll need several coats. Don't make it too thick either, or it may bubble or peel.

Beware too: many colours are affected by the alkalis in cement. Iron oxide, which gives a red colour, isn't — but do experiment.

Note: do keep a careful record of your measurements while you are experimenting! If you don't EXACTLY use the same measurements next time, you won't get the same colour.

## Victoria's interior paint

This looks stunning: a wonderfully warm creamy colour, beautifully smooth and longlasting on their rammed earth walls. Victoria says she didn't use a recipe: just bunged together dried milk that had been in the cupboard too long, ochres and some elderly yoghurt, with milk to water it down.

## Cow manure wash

No, this doesn't stink! At least not after it's been painted on. I rather like the smell of cow manure, at least as long as it's in a paddock and not in an overused barn. Again, use it only on the materials detailed above. It will only take on an unpainted porous surface, and is best as an interior paint.

1 kg hydrated lime
4 kg fresh sploshy cow manure. If it's dry you may need to double or even triple the amount!

Fresh cow manure from good green grass gives a pale green to greenish cream colour — really quite lovely. Older cow manure gives a richer creamy colour — also lovely.

Mix well, with as much water as is needed to make it creamy. Strain out ALL the lumps and bits of undigested grass. Use it at once. This paint can be improved by combining it with the basic paint recipe: use milk made from twice as much milk powder as the recipe recommends instead of water and cement instead of lime.

## Water-resistant whitewash or paint

This recipe used to be popular because all the ingredients were free if you cared to forage for them.

1 part cement
1 part rock salt
2 parts hydrated lime
4 parts powdered milk, old yoghurt or fresh milk.

Mix well till creamy. Add water if you are using powdered milk otherwise no water will be needed. Paint at once. Note: the land environment really doesn't need any more salt added to it . . . but there may be a time when you'll find this useful!

## Oil and beeswax sealers

Linseed oil comes from pressed flax seeds — you get more oil if you heat them first. Other oils, notably citrus oils, are also used as wood sealers, mostly for indoor unpainted wood. Heat a cup of beeswax till it melts. Take it off the heat and add about six cups of oil and three cups of turps, but experiment to find the thickness you prefer to use. This must be reapplied every few years and isn't suitable for wood that has previously been painted.

## Wallpaper

Wallpaper used to be popular mostly because it made rough walls smooth and the smoother your walls are, the less likely they are to accumulate dust and cobwebs. The simplest wallpaper was old newspaper: you can still find it around here glued onto wattle and daub (mud and small branch) walls, and fascinating it is too. You can read all the gossip of the 1920s and '30s. Most newspaper 'wallpaper' though was painted over.

Fabric can also be used as wallpaper: silk 'papered' walls are very, very fashionable, and expensive and they've been fashionable and expensive for a few thousand years. Linen is also used as 'wallpaper'.

If I had enough paperbarks (melaleucas) I think I'd have at least one room papered with its bark. Tear it off and paste it on, all shaggy and lovely. This does, of course, defeat the original purpose of wallpaper, but paperbark is waterproof: it's great for a dampish wall, assuming it stays glued on and would also make good insulation. And it would be almost like living in the heart of a paperbark tree.

## Paint stripper

This is NOT a safe recipe, for either you or the environment, but it is probably a bit better than a caustic soda bath. Use gloves, goggles and all protective clothing, use out of doors and NOT NEAR KIDS OR ANIMALS. You have been warned. DO NOT BLAME ME OR THE PUBLISHER FOR ACCIDENTS OR SIDE EFFECTS!

If at all possible, scrape off old paint instead (you can buy special doovers at the hardware store) then use sandpaper or a sander, or use a blow torch if it's old and thick, plus the scraper thingie.

1 kg caustic soda
1 kg plain flour
1 litre water.

Mix dry ingredients; add water CAUTIOUSLY and stir well with a wooden stake, like a tomato stake or similar, while it 'cooks' and turns to paste. Use this STRAIGHT AWAY.

Use a thick cloth, gloves and long sleeves and other protective clothing too, to smear it onto furniture or other painted wood; leave it on for 24 hours.

Scrape it off and the paint should come too.

## Homemade putty

Mix ground chalk with linseed oil till, well, putty-like. You can also add sawdust to give it more of a wood colour and texture. It's useful to fill around windows or gaps in tiles or wood. Putty can often be reused by scraping it off, removing the tough skin and heating it VERY gently with a little more linseed oil, then working the oil gently into the putty. If you burn it or yourself don't blame me.

# KEEPING COSTS DOWN AND SECONDHAND MATERIALS

Like rammed earth, stone is a very cheap building material. However, the cost of the walls is only about a fifth of the cost of your house: cheap walls still don't give you a cheap house! The most expensive bits of your house will be the foundations — something you can't skimp on; and the roof, *ditto*.

## Roofing

The best way to save on roofing is to have a tall narrow house: two storeys and an attic perhaps, with a very simple roof line. Don't be tempted to use secondhand roofing material; you'll just have to replace it sooner. Most secondhand roofing material has been damaged or distorted when it was removed, even if it was in good shape then. Especially don't use any rusty stuff! Zinc-treated roofing shouldn't be used with galvanised steel as it will corrode the galvanised sheet.

You can sometimes buy roofing tiles secondhand; always check that they're not damaged and eroded, and that you do have enough to do the entire roof! You may well not be able to get any more.

The only cheap or free roofing materials I know of are thatch, which is not really suitable for vermin and bushfire prone Australia; hand-baked clay tiles (supposing you have both the skill and equipment to make them) are another free one; wood slabs, and again you must have both the wood and the skill, and not be in a bushfire prone area; and building a domed stone, or concrete, or mudbrick roof.

I once built a roof made of old chicken wire. We then plastered newspaper over that, then covered that with a thin layer of mortar. The inside was later plastered, the outside was painted with waterproof sealer (then painted purple — but that's another story). Thirty years later it's still there, still waterproof.

**Like rammed earth, stone is a very cheap building material. However, the cost of the walls is only about a fifth of the cost of your house.**

That roof, however, was never approved by the local council. I don't know if it would or could have been, or if an engineer would have passed it. I would also have loved to build a domed roof with greater reinforcing, then covered it with stone facing . . . one day, maybe.

## Doors and windows

Check doors and windows for rot. Even if you do find an area of rot, it may be possible to replace it with a fresh piece of timber, or take the window apart and reglue and nail it if it's wobbly. One of my favourite windows in my study was half rotted, shaky, shonky and stained. But it's a lovely shape, and worth the effort to repair it. Cracked or broken glass can be replaced. (But you may never get around to it! Twenty years on one of our kitchen windows still has tape over a crack in the corner!)

I'd only bother with solid timber secondhand doors. Cheaper particle board ones don't last as long, and may not be worth buying or installing.

## Bathroom and kitchen fittings

You can usually pick up quite cheap and good secondhand kitchen sinks, baths, basins and toilets, replaced with more trendy ones. Check for chips and also rust and other stains which may be difficult to remove. Old baths can be re-enamelled; check in the *Yellow Pages*.

Don't bother with secondhand particle board cupboards and benches, unless they are very very cheap (like, free to a good home). They don't have a very long life in any case, and will have an even shorter one after they've been shoved about.

You can often buy 'seconds' from fittings manufacturers. Make sure they don't have any major defect before you stick your money down, but usually they're a pretty good buy (our toilet has

## RADIOACTIVE ROCKS

*This is a subject I don't want to know about, but you might. Some areas have higher background radiation than others. Areas with granite rock are one of these, and houses made of granite and some other rocks may greatly add to the amount of radiation you absorb. People who live in granite house in areas like Cornwall do have a higher risk of radiation-associated cancers — not a greatly increased risk, but still statistically significant.*

*I simply don't want to know how much greater the risk I have of developing cancer is because I live in a granite house: I think (seriously) the stress of knowing would be a greater health hazard than the radiation.*

*I gather (and remember this is advice from someone who has deliberately stayed ignorant — probably the only area in my life where that's so!) that houses with basements and cellars, or bathrooms and other rooms that are rarely opened are the worst hazard. Our frequently-opened Aussie houses perhaps aren't a health risk at all. There are few days, even in the heart of winter, when our house isn't opened up as much as possible for part of the day.*

*So there you are. You can prefer to ignore it, as most people ignore the far more common housing risks of formaldehyde, the vapours from many cleaning products and other noxious gasses that accumulate in houses, the possible health risks of gas heating and cooking — I could go on but I don't want to make you paranoid.*

*On the other hand you may decide to investigate all this further, and check the background radiation level of any area before you decide to live there.*

*I love this place far too deeply to even consider moving. But the choice is yours.*

a small chip out of one side. Who cares?). Plumbing fittings should mostly be new.

### Where to find cheap materials

- Secondhand building suppliers: look in the *Yellow Pages*.
- Demolition companies.
- Renovation companies: kitchen and bathroom stuff can often be bought cheaply or scavenged from companies putting new fashionable kitchens into old homes. Our first benches were third hand: pretty tatty, even with new paint, but they lasted us 15 years or so till we could afford new ones. Well, new ones made of recycled wood anyway.
- Classified ads: you may find an ad for someone who wants to demolish a garage or other structure and will swap labour for the right to take away the materials; companies may have surplus stock and want to get rid of it quickly.
- Dumps and recycling centres: sometimes you can ask the manager of the recycling centre to set certain materials aside for you to buy. Otherwise it's just a matter of visiting often.

Building a stone house will give you a stunning solid house; it will save you money. But it's only part of the process of getting a cheap house.

## The do it yourself fallacy

*(Or maybe I should spell that 'phallacy', as it seems to apply mostly to males.)*

*Don't get me wrong about this one. I am a great believer in doing things yourself . . . BUT . . . just because you CAN do something yourself, it doesn't mean you have to. At the moment I have a dear friend who has been waiting for five and a half years for a cupboard to be finished. If she attempted to do it herself her husband — who is an excellent amateur carpenter — would feel it an enormous insult to his abilities. "Of course I'll get around to it!" he cries. "But I have to finish the bathroom, chook shed (insert 15 other projects, plus a full time job and a Phd) first." Meanwhile the toddlers have to be barricaded out of the kitchen. If she — horror of horrors — actually employed someone, she feels it might be the end of their marriage.*

*Another friend is still waiting for a bathroom door ten years after the house was (almost) completed. I can also bring to mind a collapsed sofa propped up by a coffee table, holey screen doors, a ladder instead of stairs to the bedroom . . . all projects that need doing, but probably won't be tackled till they rise up the list.*

*I repeat . . . just because you CAN do something yourself, it doesn't mean you have to. Doing things yourself can become addictive; as though it brings a sense of shame and failure if you admit there are a few things you don't have time for.*

*By all means, DO do as much as you can by yourself, or with friends; but don't let it take over, so your image of yourself depends on being able to do everything, especially as you get older, or you get a crook knee, or toddlers who need more of your time.*

*Building my own house — well, some of it — was one of the most satisfying things I've ever done. I do get a great sense of confidence knowing that if an alien dropped me on another earthlike planet I could build my house from what I found there.*

*I like knowing how to dig out a road with a mattock, build a chook house, put down a concrete slab, tile a floor. But I probably won't ever do any of these again, even if (as is likely) our house continues to grow, or we build the guest studio we're planning.*

*Life is very short. There are a hundred books I still want to write, ten thousand lunches with friends I want to eat, hypothetical grandchildren I want to play with.*

*There's a time for every purpose under heaven. A time to plant, a time to build . . . and a time to ask someone else to give you a hand.*

# Ground Floor Plan

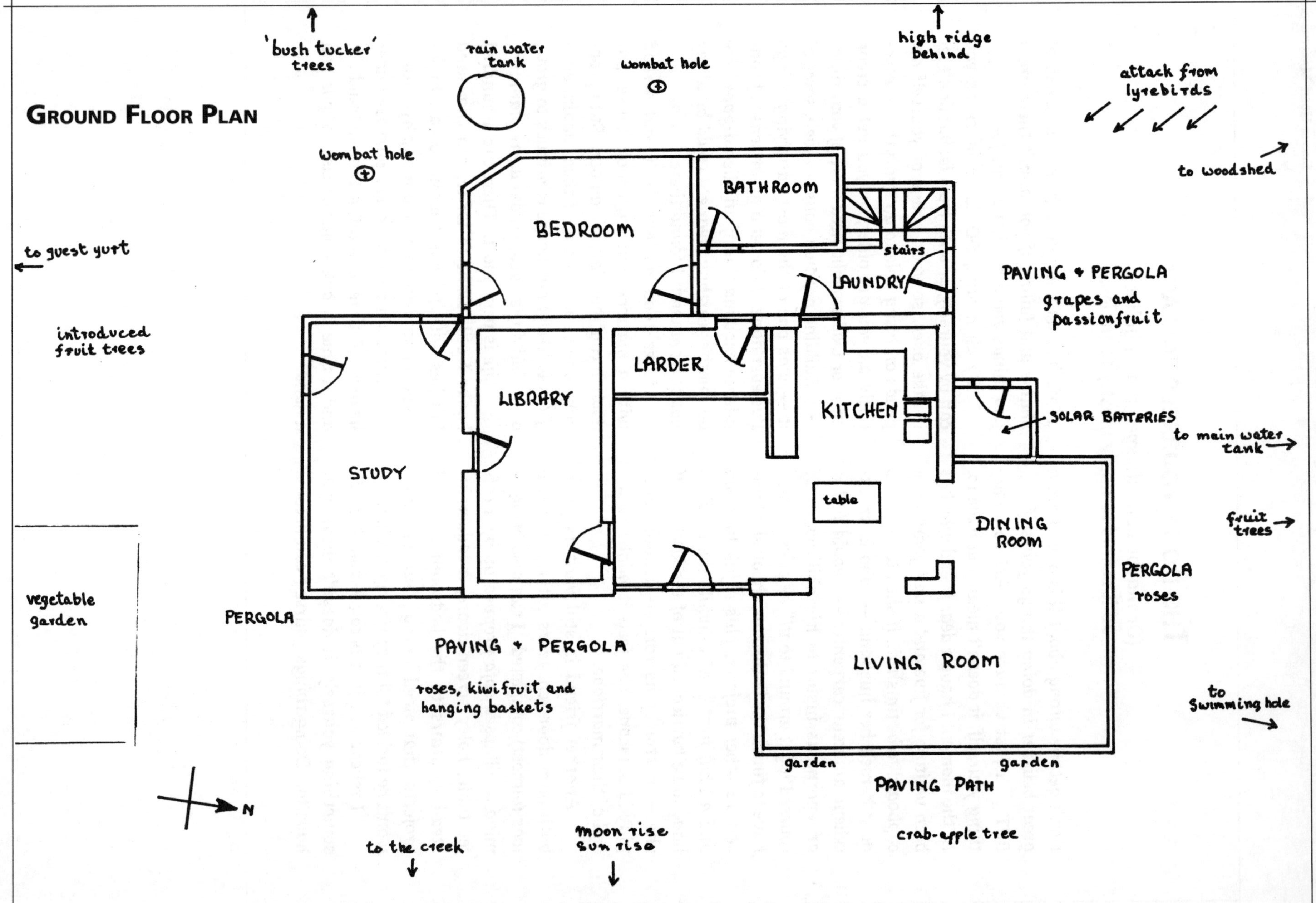

## First Floor Plan

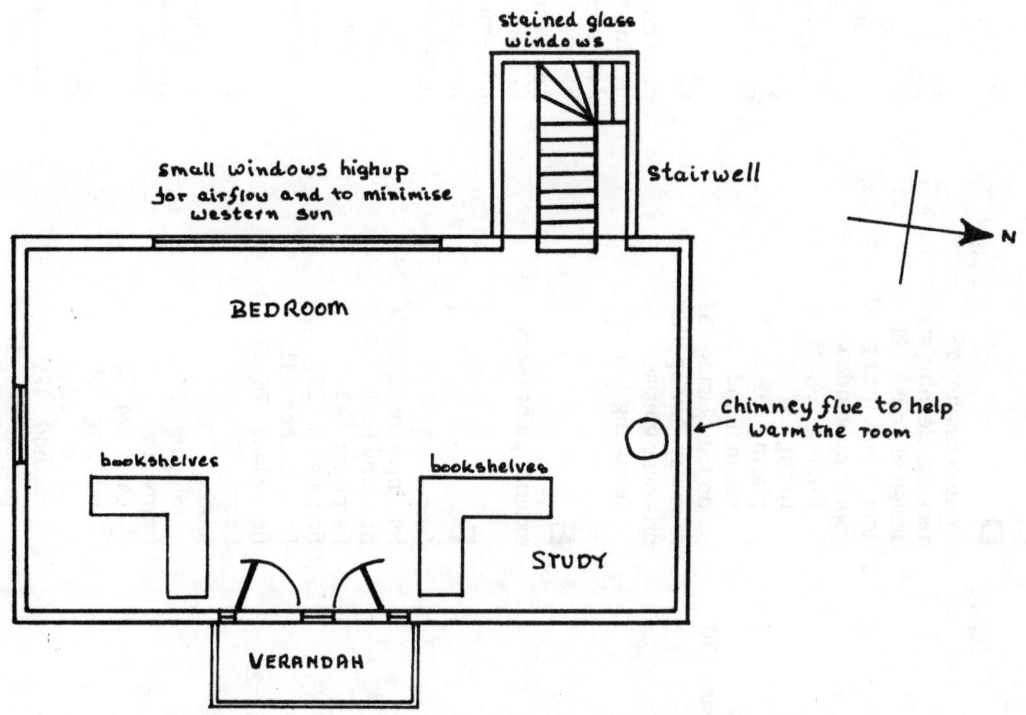

# INDEX

# Other Jackie French titles:

**These enduring, practical books written by Jackie French are available from 'The Good Life' Book Club, Australia's no-obligation book club for simpler, Greener, helathier lifestyles.**

### Jackie French's Guide to Companion Planting

$10.95 inc GST

Companion planting is really understanding how things work in your garden. This handy guide exposes popular myths and shows what works. 125 pages.

### Organic Control Of Household Pests

$14.95 inc GST

Jackie describes easy, sensible ways to control the whole range of household pests from flies, mosquitoes, ants, fleas, silverfish and cockroaches, to dust mites, cane toads, rodents and wood borers. She also gives recipes for sprays, baits, oils and repellents, lists useful plants to grow for their natural repellent powers. A must for any Green bookshelf. 138 pages.

### Making Money From Your Garden

$10.95 inc GST

This practical, down-to-earth book is brimming with ideas on how to turn your home and garden into a source of extra income. Over one hundred ideas to get you started — everything from selling surplus vegetables to making perfume. 80 pages.

### Jackie French's Chook Book

$13.95 inc GST

This handy book has lots of advice and clever ideas to manage flocks, and includes lots of chicken and egg recipes. 120 pages.

### Jackie French's Household Herb Book

$10.95 inc GST

Australia's organic garden expert has produced a readable, comprehensive and witty guide to everything important about household herbs. Learn about how to grow herbs, and how to use them for cooking, remedies, cleaning, baths, massages, aphrodisiacs, for crafts, first aid, pets and other uses. 80 pages.

### The Earth Gardener's Companion

$14.95 inc GST

Reprinted by popular demand.

This classic text has guided thousands of Australian organic gardeners and is now back in print. Read what you should be doing in the organic garden for any month of the year. Includes harvesting, planting, pest control recipes and much more. Stunning value. 80 pages, large format paperback.

### The Best of Jackie French

$29.95 inc GST

This 'encyclopedia' is the 'best of' our favourite organic gardening expert. From garden to house to table, it starts in the soil with how to plant and grow, then goes inside with practical and delightful ideas for cooking and using your homegrown produce. The book is organised A to Z and is vintage Jackie, full of her sense of fun, her passion for organics and self-sufficiency, her love of chooks, and her inspiring style of writing. A wonderful gift for yourself, or for any friend. 384 pages.

### Backyard self-Sufficiency

$17.95 inc GST

Self-sufficiency doesn't mean a huge acreage. You can use an ordinary backyard plot. Hundreds of ideas to help you create your own Garden of Eden — all detailed in the engaging style that has made Jackie a star of magazines, radio and 'Burke's Backyard'. 164 pages.

***ORDER NOW: phone (03) 5424 1814, fax (03) 5424 1743, write to PO Box 2 Trentham, Victoria, 3458, or email: <goodlife@earthgarden.com.au>.***